R Machine Learning Projects

Implement supervised, unsupervised, and reinforcement
learning techniques using R 3.5

Dr. Sunil Kumar Chinnamgari

BIRMINGHAM - MUMBAI

R Machine Learning Projects

Commissioning Editor: Sunith Shetty
Acquisition Editor: Yogesh Deokar
Content Development Editor: Atikho Sapuni Rishana
Technical Editor: Vibhuti Gawde
Copy Editor: Safis Editing
Project Coordinator: Kirti Pisat
Proofreader: Safis Editing
Indexer: Mariammal Chettiyar
Graphics: Jisha Chirayil
Production Coordinator: Tejal Daruwale Soni

First published: January 2019

Production reference: 1100119

Published by Packt Publishing Ltd.
Livery Place
35 Livery Street
Birmingham
B3 2PB, UK.

ISBN 978-1-78980-794-3

www.packtpub.com

`mapt.io`

Mapt is an online digital library that gives you full access to over 5,000 books and videos, as well as industry leading tools to help you plan your personal development and advance your career. For more information, please visit our website.

Why subscribe?

- Spend less time learning and more time coding with practical eBooks and Videos from over 4,000 industry professionals

- Improve your learning with Skill Plans built especially for you

- Get a free eBook or video every month

- Mapt is fully searchable

- Copy and paste, print, and bookmark content

Packt.com

Did you know that Packt offers eBook versions of every book published, with PDF and ePub files available? You can upgrade to the eBook version at `www.packt.com` and as a print book customer, you are entitled to a discount on the eBook copy. Get in touch with us at `customercare@packtpub.com` for more details.

At `www.packt.com`, you can also read a collection of free technical articles, sign up for a range of free newsletters, and receive exclusive discounts and offers on Packt books and eBooks.

Dedicated to my loving wife, HimaBindu. Sometimes I wonder what am I without you; I've never told you this, but you actually define me.

I'd like to thank my dear friend Nanditha Siva for being there by my side in the difficult times. You are the one who has always had faith in me that "I am something".

Contributors

About the author

Dr. Sunil Kumar Chinnamgari has a PhD in computer science (specializing in machine learning and natural language processing). He is an AI researcher with more than 14 years of industry experience. Currently, he works in the capacity of a lead data scientist with a US financial giant. He has published several research papers in Scopus and IEEE journals, and is a frequent speaker at various meet-ups. He is an avid coder and has won multiple hackathons. In his spare time, Sunil likes to teach, travel, and spend time with family.

About the reviewers

Davor Lozić is a senior software engineer interested in various subjects, especially computer security, algorithms, and data structures. He manages teams of more than 15 engineers and is a professor teaching about database systems. You can contact him at davor@warriorkitty.com. He likes cats! If you want to talk about any aspect of technology, or if you have funny pictures of cats, feel free to contact him.

Giuseppe Ciaburro holds a PhD in environmental technical physics and two master's degrees. His research focuses on machine learning applications in the study of urban sound environments. He works at Built Environment Control Laboratory—Università degli Studi della Campania Luigi Vanvitelli (Italy). He has over 15 years of work experience in programming (in Python, R, and MATLAB), first in the field of combustion and then in acoustics and noise control. He has several publications to his credit.

Packt is searching for authors like you

If you're interested in becoming an author for Packt, please visit authors.packtpub.com and apply today. We have worked with thousands of developers and tech professionals, just like you, to help them share their insight with the global tech community. You can make a general application, apply for a specific hot topic that we are recruiting an author for, or submit your own idea.

Table of Contents

Preface

R is one of the most popular languages when it comes to performing computational statistics (statistical computing) easily and exploring the mathematical side of machine learning. With this book, you will leverage the R ecosystem to build efficient machine learning applications that carry out intelligent tasks within your organization.

This book will help you test your knowledge and skills, guiding you on how to build easy through to complex machine learning projects. You will first learn how to build powerful machine learning models with ensembles to predict employee attrition. Next, you'll implement a joke recommendation engine to perform sentiment analysis on Amazon reviews. You'll also explore different clustering techniques to segment customers using wholesale data. In addition to this, the book will get you acquainted with credit card fraud detection using autoencoders, and reinforcement learning to make predictions and win on a casino slot machine.

By the end of the book, you will be equipped to confidently perform complex tasks to build research and commercial projects for automated operations.

Who this book is for

This book is for data analysts, data scientists, and ML developers who wish to master the concepts of ML using R by building real-world projects. Each project will help you test your expertise to implement the working mechanisms of ML algorithms and techniques. A basic understanding of ML and a working knowledge of R programming is a must.

What this book covers

Chapter 1, *Exploring the Machine Learning Landscape*, will briefly review the various ML concepts that a practitioner must know. In this chapter, we will cover topics such as supervised learning, reinforcement learning, unsupervised learning, and real-world ML uses cases.

Chapter 2, *Predicting Employee Attrition Using Ensemble Models*, covers the creation of powerful ML models through ensemble learning. The project covered in this chapter is from the human resources domain. Retention of talented employees is a key challenge faced by corporations. If we were able to predict the attrition of an employee well in advance, it is possible that the human resources or management team could do something to save the potential attrition from becoming real. It just so happens that it is possible to predict employee attrition through the application of ML. This chapter makes use of an IBM-curated public dataset that provides a pseudo employee attrition population and characteristics. We start the chapter with an introduction to the problem at hand and then attempt to explore the dataset with **exploratory data analysis** (**EDA**). The next step is the preprocessing phase, which includes the creation of new features using prior domain experience. Once the dataset is fully prepared, models will be created using multiple ensemble techniques, such as bagging, boosting , stacking, and randomization. Lastly, we will deploy the finally selected model for production. We will also learn about the concepts underlying the various ensemble techniques used to create the models.

Chapter 3, *Implementing a Joke Recommendation Engine*, introduces recommendation engines, which are designed to predict the ratings that a user would give to content such as movies and music. Based on what a user has previously liked or seen and using other profiling attributes, a recommendation engine suggests new content that the user might like. Such engines have gained a lot of significance in recent years. We explore the exciting area of recommendation systems by working on a joke recommendation engine project. In this chapter, we start by understanding the concepts and types of collaborative filtering algorithms. We will then build a recommendation engine to provide personalized joke recommendations using collaborative filtering approaches such as user-based collaborative filters and item-based collaborative filters. The dataset used for this project is a open dataset called the Jester jokes dataset. Apart from this, we will be exploring various libraries available in R that can be used to build recommendation systems, and we will be comparing the performances obtained from these approaches. Additionally, we leverage the market basket analysis technique, a pretty popular technique in the marketing domain, to discern relationships between various jokes.

Chapter 4, *Sentiment Analysis of Amazon Reviews with NLP*, covers sentiment analysis, which entails finding the sentiment of a sentence and labeling it as positive, negative, or neutral. This chapter introduces sentiment analysis and covers the various techniques that can be used to analyze text. We will understand text-mining concepts and the various ways that text is labeled based on the tone.

We will apply sentiment analysis to Amazon product review data. This dataset contains millions of Amazon customer reviews and star ratings. It is a classification task where we will be categorizing each review as positive, negative, or neutral depending on the tone. Apart from using various popular R text-mining libraries to preprocess the reviews to be classified, we will also be leveraging a wide range of text representations, such as bag of words, word2vec, fastText, and Glove. Each of the text representations is then used as input for ML algorithms to perform classification. In the course of implementing each of these techniques, we will also learn about the concepts behind these techniques and also explore other instances where we could successfully apply them.

Chapter 5, *Customer Segmentation Using Wholesale Data*, covers the segmentation, grouping, or clustering of customers, which can be achieved through unsupervised learning. We explore the various aspects of customer grouping in this chapter. Customer segmentation is an important tool used by product sellers to understand their customers and gather information. Customers can be segmented based on different criteria, such as age and spending patterns. In this chapter, we learn the various techniques of customer segmentation. For the project, we use a dataset containing wholesale transactions. This dataset is available in the UCI Machine Learning Repository. We will be applying advanced clustering techniques, such as k-means, DIANA, and AGNES. At times, we will not know the number of groups that exist in the dataset at hand. We will explore the ML techniques for dealing with such ambiguity and have ML find out the number of groups possible based on the underlying characteristics of the input data. Evaluating the output of the clustering algorithms is an area that is often challenging to practitioners. We also explore this area so as to have a well-rounded understanding of applying clustering algorithms to real-world problems.

Chapter 6, *Image Recognition Using Deep Neural Networks*, covers **convolutional neural networks (CNNs)**, which are a type of deep neural network and are popular in computer vision applications. In this chapter, we learn about the fundamental concepts underlying CNNs. We explore why CNNs work so well with computer vision problems such as object detection. We discuss the aspects of transfer learning and how it works in tandem with CNNs to solve computer vision problems. As elsewhere in the book, we'll be going by the philosophy of learning by doing. We will learn about all of these concepts by applying a CNN in the building of a multi-class classification model on a popular open dataset called MNIST. The objective of the project is to classify given images of handwritten digits. The project explores the methodology for creating features from raw images. We will learn about the various preprocessing techniques that can be applied to the image data in order use the data with deep learning models.

Chapter 7, *Credit Card Fraud Detection Using Autoencoders*, covers autoencoders, which are yet another type of unsupervised deep learning network. We start the chapter by understanding autoencoders and how they are different from the other deep learning networks, such as **recurrent neural networks** (**RNNs**)and CNNs. We will learn about autoencoders by implementing a project that identifies credit card fraud. Credit card companies are constantly seeking ways to detect credit card fraud. Fraud detection is a key aspect for banks to protect their revenues. It can be achieved through the application of ML in the finance domain for the specific fraud detection problem. A fraud is usually an anomalous event that requires immediate action. In this chapter, we will use an autoencoder to detect fraud. Autoencoders are neural networks that contain a bottleneck layer whose dimensionality is smaller than the input data. In this chapter, we will become familiar with dimensionality reduction and how it can be used to identify credit card fraud detection. For the project, we will be using the H2O deep learning framework in tandem with R. As far as the dataset is concerned, we use an open dataset that contains credit card transactions of European card holders from September 2013. There are a total of 284,807 transactions, out of which 492 are fraudulent.

Chapter 8, *Automatic Prose Generation with Recurrent Neural Networks*, introduces some **deep neural networks** (**DNNs**) that have recently received a lot of attention. This is due to their success in obtaining great results in various areas of ML, from face recognition and object detection to music generation and neural art. This chapter introduces the concepts necessary for understanding deep learning. We discuss the nuts and bolts of neural networks, such as neurons, hidden layers, various activation functions, techniques for dealing with problems faced in neural networks, and using optimization algorithms to get weights in neural networks. We will also implement a neural network from scratch to demonstrate these concepts. The content of this chapter will help us get foundational knowledge on neural networks. Then, we will learn how to apply an RNN by doing a project. It has always been thought that creative tasks such as authoring stories, writing poems, and painting pictures can only be achieved by humans. This is no longer true, thanks to deep learning! Technology can now accomplish creative tasks. We will create an application based on **long short-term memory** (**LSTM**) network, a variant of RNNs that generates text automatically. To accomplish this task, we make use of the MXNet framework, which extends its support for the R language to perform deep learning. In the course of implementing this project, we will also learn more about the concepts surrounding RNNs and LSTMs.

Chapter 9, *Winning the Casino Slot Machines with Reinforcement Learning*, begins with an explanation of RL. We discuss the various concepts of RL, including strategies for solving what is called as the multi-arm bandit problem. We implement a project that uses UCB and Thompson sampling techniques in order to solve the multi-arm bandit problem.

`Appendix`, *The Road Ahead*, briefly discuss the advancements in the ML world and the need to stay on top of them.

To get the most out of this book

The projects covered in this book are intended to expose you to practical knowledge on the implementation of various ML techniques to real-world problems. It is expected that you have a good working knowledge of R and some basic understanding of ML. Basic knowledge of ML and R is a must prior to starting this project.

It should also be noted that the code for the projects is implemented using R version 3.5.2 (2018-12-20), nicknamed Eggshell Igloo. The project code has been successfully tested on Linux Mint 18.3 Sylvia. There is no reason to believe that the code does not work on other platforms, such as Windows; however, this is not something that has been tested by the author.

Download the example code files

You can download the example code files for this book from your account at www.packt.com. If you purchased this book elsewhere, you can visit www.packt.com/support and register to have the files emailed directly to you.

You can download the code files by following these steps:

1. Log in or register at www.packt.com.
2. Select the **SUPPORT** tab.
3. Click on **Code Downloads & Errata**.
4. Enter the name of the book in the **Search** box and follow the onscreen instructions.

Once the file is downloaded, please make sure that you unzip or extract the folder using the latest version of:

- WinRAR/7-Zip for Windows
- Zipeg/iZip/UnRarX for Mac
- 7-Zip/PeaZip for Linux

The code bundle for the book is also hosted on GitHub at `https://github.com/PacktPublishing/R-Machine-Learning-Projects`. In case there's an update to the code, it will be updated on the existing GitHub repository.

We also have other code bundles from our rich catalog of books and videos available at `https://github.com/PacktPublishing/`. Check them out!

Download the color images

We also provide a PDF file that has color images of the screenshots/diagrams used in this book. You can download it here: `http://www.packtpub.com/sites/default/files/downloads/9781789807943_ColorImages.pdf`.

Conventions used

There are a number of text conventions used throughout this book.

`CodeInText`: Indicates code words in text, database table names, folder names, filenames, file extensions, pathnames, dummy URLs, user input, and Twitter handles. Here is an example: "The `rsample` library incorporates this dataset, and we can make use of this dataset directly from the library."

A block of code is set as follows:

```
setwd("~/Desktop/chapter 2")
library(rsample)
data(attrition)
str(attrition)
mydata<-attrition
```

When we wish to draw your attention to a particular part of a code block, the relevant lines or items are set in bold:

```
[default]
exten => s,1,Dial(Zap/1|30)
exten => s,2,Voicemail(u100)
exten => s,102,Voicemail(b100)
exten => i,1,Voicemail(s0)
```

Bold: Indicates a new term, an important word, or words that you see onscreen. For example, words in menus or dialog boxes appear in the text like this. Here is an example: "You may recollect the **Customers Who Bought This Item Also Bought This** heading on Amazon (or any e-commerce site) where recommendations are shown."

Warnings or important notes appear like this.

 Tips and tricks appear like this.

Get in touch

Feedback from our readers is always welcome.

General feedback: If you have questions about any aspect of this book, mention the book title in the subject of your message and email us at customercare@packtpub.com.

Errata: Although we have taken every care to ensure the accuracy of our content, mistakes do happen. If you have found a mistake in this book, we would be grateful if you would report this to us. Please visit www.packt.com/submit-errata, selecting your book, clicking on the Errata Submission Form link, and entering the details.

Piracy: If you come across any illegal copies of our works in any form on the Internet, we would be grateful if you would provide us with the location address or website name. Please contact us at copyright@packt.com with a link to the material.

If you are interested in becoming an author: If there is a topic that you have expertise in and you are interested in either writing or contributing to a book, please visit authors.packtpub.com.

Reviews

Please leave a review. Once you have read and used this book, why not leave a review on the site that you purchased it from? Potential readers can then see and use your unbiased opinion to make purchase decisions, we at Packt can understand what you think about our products, and our authors can see your feedback on their book. Thank you!

For more information about Packt, please visit packt.com.

Exploring the Machine Learning Landscape

1

Machine learning (**ML**) is an amazing subfield of **Artificial Intelligence** (**AI**) that tries to mimic the learning behavior of humans. Similar to the way a baby learns by observing the examples it encounters, an ML algorithm learns the outcome or response to a future incident by observing the data points that are provided as input to it.

In this chapter, we will cover the following topics:

- ML versus software engineering
- Types of ML methods
- ML terminology—a quick review
- ML project pipeline
- Learning paradigm
- Datasets

ML versus software engineering

With most people transitioning from traditional software engineering practice to ML, it is important to understand the underlying difference between both areas. Superficially, both of these areas seem to generate some sort of code to perform a particular task. An interesting fact to observe is that, unlike software engineering where a programmer explicitly writes a program with various responses based on several conditions, the ML algorithm infers the rules of the game by observing the input examples. The rules that are learned are further used for better decision making when new input data is fed to the system.

As you can observe in the following diagram, automatically inferring the actions from data without manual intervention is the key differentiator between ML and traditional programming:

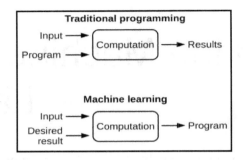

Another key differentiator of ML from traditional programming is that the knowledge acquired through ML is able to generalize beyond the training samples by successfully interpreting data that the algorithm has never seen before, while a program coded in traditional programming can only perform the responses that were included as part of the code.

Yet another differentiator is that in software engineering, there are certain specific ways to solve a problem at hand. Given an algorithm developed based on certain assumptions of inputs and the conditions incorporated, you will be able to guarantee the output that will be obtained given an input. In the ML world, it is not possible to provide such assurances on the output obtained from the algorithms. It is also very difficult in the ML world to confirm if a particular technique is better than another without actually trying both the techniques on the dataset for the problem at hand.

 ML and software engineering are not the same! ML projects may involve some software engineering in them, but ML cannot be considered to be the same as software engineering.

While there is more than one formal definition that exists for ML, the following mentioned are a few key definitions encountered often:

"Machine learning is the science of getting computers to act without being explicitly programmed."

—Stanford

"Machine learning is based on algorithms that can learn from data without relying on rules-based programming."

—McKinsey and Co.

With the rise of data as the fuel of the future, the terms AI, ML, data mining, data science, and data analytics are used interchangeably by industry practitioners. It is important to understand the key differences between these terms to avoid confusion.

 The terms AI, ML, data mining, data science, and data analytics, though used interchangeably, are not the same!

Let's take a look at the following terms:

- **AI**: AI is a paradigm where machines are able to perform tasks in a smart way. It may be observed that in the definition of AI, it is not specified whether the smartness of machines may be achieved manually or automatically. Therefore, it is safe to assume that even a program written with several `if...else` or `switch...case` statements that has then been infused with a machine to carry out tasks may be considered to be AI.
- **ML**: ML, on the other hand, is a way for the machine to achieve smartness by learning from the data that is provided as input and, thereby, we have a smart machine performing a task. It may be observed that ML achieves the same objective of AI except that the smartness is achieved automatically. Therefore, it can be concluded that ML is simply a way to achieve AI.
- **Data mining**: Data mining is a specific field that focuses on discovering the unknown properties of the datasets. The primary objective of data mining is to extract rules from large amounts of data provided as input, whereas in ML, an algorithm not only infers rules from the data input, but also uses the rules to perform predictions on any new, incoming data.
- **Data analytics**: Data analytics is a field that encompasses performing fundamental descriptive statistics, data visualization, and data points communication for conclusions. Data analytics may be considered to be a basic level within data science. It is normal for practitioners to perform data analytics on the input data provided for data mining or ML exercises. Such analysis on data is generally termed as **exploratory data analysis (EDA)**.

- **Data science**: Data science is an umbrella term that includes data analytics, data mining, ML, and any specific domain expertise pertaining to the field of work. Data science is a concept that includes several aspects of handling the data such as acquiring the data from one or more sources, data cleansing, data preparation, and creating new data points based on existing data. It includes performing data analytics. It also encompasses using one or more data mining or ML techniques on the data to infer knowledge to create an algorithm that performs a task on unseen data. This concept also includes deploying the algorithm in a way that it is useful to perform the designated tasks in the future.

The following is a Venn diagram which demonstrates the skills required by a professional working in the data science ambit. It has three circles, each of which defines a specific skill that a data science professional should have:

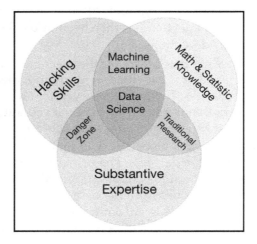

Let's explore the following skills mentioned in the preceding diagram:

- **Math & Statistic Knowledge**: This skill is required to analyze the statistical properties of the data.
- **Hacking Skills**: Programming skills play a key role in order to process the data in a quick manner. The ML algorithm is applied to create an output that will perform the prediction on unseen data.
- **Substantive Expertise**: This skill refers to the domain expertise in the field of the problem at hand. It helps the professional to be able to provide proper inputs to the system from which it can learn and to assess the appropriateness of the inputs and results obtained.

To be a successful data science professional you need to have math, programming skills, as well as knowledge of the business domain.

As we can see, AI, data science, data analytics, data mining, and ML are all interlinked. All of these areas are the most in-demand domains in the industry right now. The right skill sets in combination with real-world experience will lead to a strong career in these areas which are currently trending. As ML forms the core of the leading space, the next section explores the various types of ML methods that may be applied to several real-world problems.

ML is everywhere! Most of the time, we may be using something that is ML-based but don't realize its existence or the influence that it has on our lives! Let's explore together some very popular devices or applications that we experience on a daily basis, which are powered by ML:

- **Virtual personal assistants** (**VPAs**) such as **Google Allo**, **Alexa**, **Google Now**, **Google Home**, **Siri**, and so on
- Smart maps that show you traffic predictions, given your source and destination
- Demand-based price surging in Uber or similar transportation services
- Automated video surveillance in airports, railway stations, and other public places
- Face recognition of individuals in pictures posted on social media sites such as Facebook
- Personalized news feeds served to you on Facebook
- Advertisements served to you on YouTube
- **People you may know** suggestions on Facebook and other similar sites
- Job recommendations on LinkedIn, based on your profile
- Automated responses on Google Mail
- Chatbots that you converse with in online customer support forums
- Search engine results filtering
- Email spam filtering

Of course, the list does not end here. The preceding applications mentioned are just a few of the basic ones that illustrate the influence that ML has on our lives today. It is not astonishing to quote that there is no subject area that ML has not touched!

The topics in this section are by no means an exhaustive description of ML, but just a quick touch point to get us started on a journey of exploration. Now that we have a basic understanding of what ML is and where it can be applied, let's delve deeper into other ML-related topics in the next section.

Types of ML methods

Several types of tasks that aim at solving real-world problems can be achieved thanks to ML. An ML method generally means a group of specific types of algorithms that are suitable for solving a particular kind of problem and the method addresses any constraints that the problem brings along with it. For example, a constraint of a particular problem could be the availability of labeled data that can be provided as input to the learning algorithm.

Essentially, the popular ML methods are supervised learning, unsupervised learning, semi-supervised learning, reinforcement learning, and transfer learning. The rest of this section details each of these methods.

Supervised learning

A supervised learning algorithm is applied when one is very clear about the result that needs to be achieved from a problem, however one is unsure about the relationships between the data that affects the output. We would like the ML algorithm that we apply on the data to perceive these relationships between different data elements so as to achieve the desired output.

The concept can be better explained with an example—at a bank, prior to extending a loan, they would like to predict if a loan applicant would pay the loan back. In this case, the problem is very clear. If a loan is extended to a prospective customer X, there are two possibilities: that X would successfully repay the loan or X would not repay the loan. The bank would like to use ML to identify the category into which customer X falls; that is, a successful repayer of the loan or a repayment defaulter.

While the problem definition that is to be solved is clear, please note that the features of a customer that will contribute to successful loan repayment or non-repayment are not clear and this is something we would like the ML algorithm to learn by observing the patterns in the data.

The major challenge here is that we need to provide input data that represents both customers that repaid their loans successfully and also customers that failed to repay. The bank can simply look at the historical data to get the records of customers in both categories and then label each record as paid or unpaid categories as appropriate.

The records, thus labeled, now become input to a supervised learning algorithm so that it can learn the patterns of both categories of customers. The process of learning from the labeled data is called **training** and the output obtained (algorithm) from the learning process is called a **model**. Ideally, the bank would keep some part of the labeled data aside from training data so as to be able to test the model created, and this data is termed as **test data**. It should be no surprise that the labeled data that is used for training the model is called **training data**.

Once the model has been built, measurements are obtained by testing the model with test data to ensure the model yields a satisfactory level of performance, otherwise model-building iterations are carried out until the desired model performance is obtained. The model that achieved the desired performance on test data can be used by the bank to infer if any new loan applicant will be a future defaulter at all and, if so, make a better decision in terms of extending a loan to that applicant.

In a nutshell, supervised ML algorithms are employed when the objective is very clear and labeled data is available as input for the algorithm to learn the patterns from. The following diagram summarizes the supervised learning process:

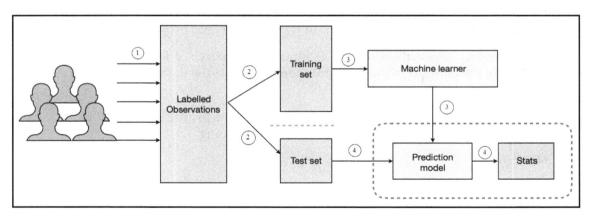

Supervised learning can be further divided into two categories, namely **classification** and **regression**. The prediction of a bank loan defaulter explained in this section is an example of classification and it aims to predict a label of a nominal type such as yes or no. On the other hand, it is also possible to predict numeric values (continuous values) and this type of prediction is called regression. An example of regression is predicting the monthly rental of a home in a prime location of a city based on features such as the demand for houses in the area, the number of bedrooms, the dimensions of the house, and accessibility to public transportation.

Several supervised learning algorithms exist, and a few popularly known algorithms in this area include **classification and regression trees** (**CART**), logistic regression, linear regression, Naive Bayes, neural networks, **k-nearest neighbors** (**KNN**), and **support vector machine** (**SVM**).

Unsupervised learning

The availability of labeled data is not very common and manually labeling data is also not cheap. This is the situation where unsupervised learning comes into play.

For example, one small boutique firm wants to roll out a promotion to its customers, who are registered on their Facebook page. While the business objective is clear—that a promotion needs to be rolled out to customers—it is unclear as to which customer falls under which group. Unlike the supervised learning method where prior knowledge existed in terms of bad debtors and good debtors, in this case there are no such clues.

When the customer information is given as input to unsupervised learning algorithms, it tries to identify the patterns in the data and thereby groups the data of the customers with similar kinds of attributes.

 Birds of the same feather flock together is the principle followed in customer grouping with unsupervised learning.

The reasoning behind the formation of these organic groups from the grouping exercise may not be very intuitive. It may take some research to identify the factors that contributed to the gathering of a set of customers in a group. Most of the time, this research is manual and the data points in each group need verifying. This research may form the basis to determine the groups to which the particular promotion at hand needs to be rolled out. This application of unsupervised learning is called **clustering**. The following diagram shows the application of unsupervised ML to cluster the data points:

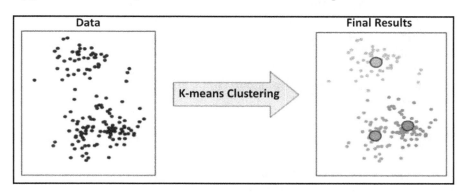

There are a number of clustering algorithms. However, the most popular ones are namely, k-means clustering, k-modes clustering, hierarchical clustering, fuzzy clustering, and so on.

Other forms of unsupervised learning do exist. For example, in retail industry, an unsupervised learning method called **association rule mining** is applied on customer purchases to identify the goods that are purchased together. In this case, unlike supervised learning, there is no need for labels at all. The task involved only requires the ML algorithm to identify the latent associations between the products that are billed together by customers. Having the information from association rule mining helps retailers place the products that are bought together in proximity. The idea is that customers can be intuitively encouraged to buy the extra products.

A priori, **equivalence class transformation** (Eclat), and **frequency pattern growth** (FPG) are popular among the several algorithms that exist to perform association rule mining.

Yet another form of unsupervised learning is anomaly detection or outlier detection. The goal of the exercise is to identify data points that do not belong to the rest of the elements that are given as input to the unsupervised learning algorithm. Similar to association rule mining, due to the nature of the problem at hand, there is no requirement for labels to be made use of by the algorithm to achieve the goal.

Fraud detection is an important application of anomaly detection in the credit cards industry. Credit card transactions are monitored in real time and any spurious transaction patterns are flagged immediately to avoid losses to the credit card user as well as the credit card provider. The unusual pattern that is monitored for could be a huge transaction in a foreign currency rather than that of a normal currency in which the particular customer generally transacts. It could be transactions in physical stores located in two different continents on the same day. The general idea is to be able to flag up a pattern that is a deviation from the norm.

K-means clustering and one-class SVM are two well-known unsupervised ML algorithms that are used to observe abnormalities in the population.

Overall, it may be understood that unsupervised learning is unarguably a very important method, given that labeled data used for training is a scarce resource.

Semi-supervised learning

Semi-supervised learning is a hybrid of both supervised and unsupervised methods. ML requires large amounts of data for training. Most of the time, a directly proportional relationship is observed between the amount of data used for model training and the performance of the model.

In niche domains such as medical imagining, a large amount of image data (MRIs, x-rays, CT scans) is available. However, the time and availability of qualified radiologists to label these images is scarce. In this situation, we might end up getting only a few images labeled by radiologists.

Semi-supervised learning takes advantage of the few labeled images by building an initial model that is used to label the large amount of unlabeled data that exists in the domain. Once the large amount of labeled data is available, a supervised ML algorithm may be used to train and create a final model that is used for prediction tasks on the unseen data. The following diagram illustrates the steps involved in semi-supervised learning:

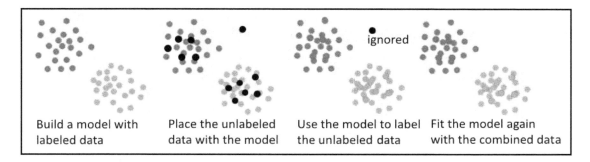

| Build a model with labeled data | Place the unlabeled data with the model | Use the model to label the unlabeled data | Fit the model again with the combined data |

Speech analysis, protein synthesis, and web content classifications are certain areas where large amounts of unlabeled data and fewer amounts of labeled data are available. Semi-supervised learning is applied in these areas with successful results.

Generative adversarial networks (GANs), **semi-supervised support vector machines (S3VMs)**, graph-based methods, and **Markov** chain methods are well-known methods among others in the semi-supervised ML area.

Reinforcement learning

Reinforcement learning (RL) is an ML method that is neither supervised learning nor unsupervised learning. In this method, a reward definition is provided as input to this kind of a learning algorithm at the start. As the algorithm is not provided with labeled data for training, this type of learning algorithm cannot be categorized as supervised learning. On the other hand, it is not categorized as unsupervised learning, as the algorithm is fed with information on reward definition that guides the algorithm through taking the steps to solve the problem at hand.

Reinforcement learning aims to improve the strategies used to solve any problem continuously by relying on the feedback received. The goal is to maximize the rewards while taking steps to solve the problem. The rewards obtained are computed by the algorithm itself going by the rewards and penalty definitions. The idea is to achieve optimal steps that maximize the rewards to solve the problem at hand.

The following diagram is an illustration depicting a robot automatically determining the ideal behavior through a reinforcement learning method within the specific context of fire:

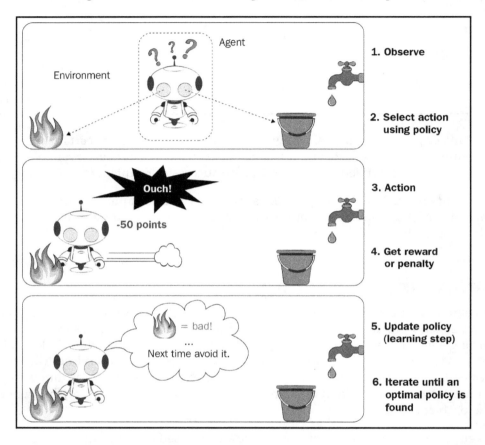

A machine outplaying humans in an Atari video game is termed as one of the foremost success stories of reinforcement learning. To achieve this feat, a large number of example games played by humans are fed as input to the algorithm that learned the steps to take to maximize the reward. The reward in this case is the final score. The algorithm, post learning from the example inputs, just simulated the pattern at each step of the game that eventually maximized the score obtained.

Though it might appear that reinforcement learning can be applied to game scenarios only, there are numerous use cases for this method in industry as well. The following examples mentioned are three such use cases:

- Dynamic pricing of goods and services based on spontaneous supply and demand targeted at achieving profit maximization is achieved through a variant of reinforcement learning called **Q-learning**.
- Effective use of space in warehouses is a key challenge faced by inventory management professionals. Market demand fluctuations, the large availability of inventory stocks, and delays in refilling the inventory are the key constraints that affect space utilization. Reinforcement learning algorithms are used to optimize the time to procure inventory as well as to reduce the time to retrieve the goods from warehouses, thereby directly impacting the space management issue referred to as a problem in the inventory management area.
- Prolonged treatments and differential drug administration is required in medical science to treat diseases such as cancer. The treatments are highly personalized, based on the characteristics of the patient. Treatment often involves variations of the treatment strategy at various stages. This kind of treatment plan is typically referred to as a **dynamic treatment regime** (DTR). Reinforcement learning helps with processing the clinical trials data to come up with the appropriate personalized DTR for the patient, based on the characteristics of the patient that are fed in as inputs to the reinforcement learning algorithm.

There are four very popular reinforcement learning algorithms, namely Q-learning, **state-action-reward-state-action** (SARSA), **deep Q network** (DQN), and **deep deterministic policy gradient** (DDPG).

Transfer learning

The reusability of code is one of the fundamental concepts of **object-oriented programming** (OOP) and it is pretty popular in the software-engineering world. Similarly, transfer learning involves reusing a model built to achieve a specific task to solve another related task.

It is understandable that to achieve better performance measurements, ML models need to be trained on large amounts of labeled data. The availability of fewer amounts of data means less training and the result is a model with suboptimal performance.

Transfer learning attempts to solve the problems arising from the availability of fewer amounts of data by reusing the knowledge obtained by a different related model. Having fewer data points available to train a model should not impede building a better model, which is the core concept behind transfer learning. The following diagram is an illustration showing the purpose of transfer learning in an image recognition task that classifies dog and cat images:

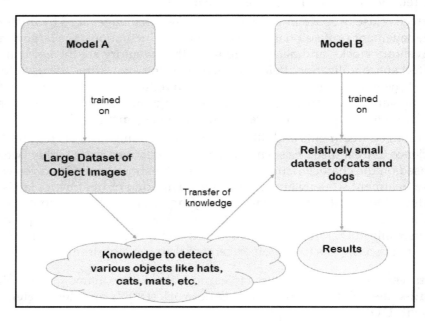

In this task, a neural network model is involved with detecting the edges, color blob detection, and so on in the first few layers. Only at the progressive layers (maybe in the last few layers) does the model attempt to identify the facial features of dogs or cats in order to classify them as one of the targets (a dog or a cat).

It may be observed that the tasks of identifying edges and color blobs are not specific to cats' and dogs' images. The knowledge to infer edges or color blobs may be generally inferred even if a model is trained on non-dog or non-cat images. Eventually, if this knowledge is clubbed with knowledge derived from inferring cat faces versus dog faces, even if they are small in number, we will have a better model than the suboptimal model obtained by training on fewer images.

In the case of a dogs-cats classifier, first, a model is trained on a large set of images that are not confined to cats' and dogs' images. The model is then taken and the last few layers are retrained on the dogs' and cats' faces. The model, thus obtained, is then tested and used post evidencing performance measurements that are satisfactory.

The concept of transfer learning is used not just for image-related tasks. Another example of it being used is in **natural language processing** (**NLP**) where it can perform sentiment analysis on text data.

Assume a company that launched a new product has a concept that never existed before (say, for now, a flying car). The task is to analyze the tweets related to the new product and identify each of them as being of positive, negative, or neural sentiment. It may be observed that prior, labeled tweets are unavailable in the flying car's domain. In such cases, we can take a model built based on the labeled data of generic product reviews for several products and domains. We can reuse the model by supplementing it with flying-car-domain-specific terminology to avail a new model. This new model will be finally used for testing and deploying to analyze sentiment on the tweets obtained about the newly launched flying cars.

It is possible to achieve transfer learning through the following two ways:

- By reusing one's own model
- By reusing a pretrained model

Pretrained models are models built by various organizations or individuals as part of their research work or as part of a competition. These models are generally very complex and are trained on large amounts of data. They are also optimized to perform their tasks with high precision. These models may take days or weeks to train on modern hardware. Organizations or individuals often release these models under permissive license for reuse. Such pretrained models can be downloaded and reused through the transfer-learning paradigm. This will effectively make use of the vast existing knowledge that the pretrained models possess, which would otherwise be hard to attain for an individual with limited hardware resources and amounts of data to train.

There are several pretrained models made available by various parties. The following described are some of the popular pretrained models:

- **Inception-V3 model**: This model has been trained on ImageNet as part of a large visual recognition challenge. The competition required the participants to classify a given image into one of 1,000 classes. Some of the classes include the names of animals and object names.

- **MobileNet**: This pretrained model has been built by Google and it is meant to perform object detection using the ImageNet database. The architecture is designed for mobiles.
- **VCG Face**: This is a pretrained model built for face recognition.
- **VCG 16**: This is a pretrained model trained on the **MS COCO** dataset. This one accomplishes image captioning; that is, given an input image, it generates a caption describing the image's contents.
- **Google's Word2Vec model and Stanford's GloVe model**: These pretrained models take text as input and produce word vectors as output. Distributed word vectors offer one form of representing documents for NLP or ML applications.

Now that we have a basic understanding of various possible ML methods, in the next section, we focus on quickly reviewing the key terminology used in ML.

ML terminology – a quick review

In this section, we take the popular ML terms and review them. This non-exhaustive review will helps us as a quick refresher and enable us to follow the projects covered by this book without any hiccups.

Deep learning

This is a revolutionary trend and has become a super-hot topic in recent times in the ML world. It is a category of ML algorithms that use **artificial neural networks (ANNs)** with multiple hidden layers of neurons to address problems.

Superior results are obtained by applying deep learning to several real-world problems. **Convolutional neural networks (CNNs)**, **recurrent neural networks (RNNs)** **autoencoders (AEs)**, **generative adversarial networks (GANs)**, and **deep belief networks (DBNs)** are some of the popular deep learning methods.

Big data

The term refers to large volumes of data that combine both structured data types (rows and columns similar to a table) and unstructured data types (text documents, voice recordings, image data, and so on). Due to the volume of data, it does not fit into the main memory of the hardware where ML algorithms need to be executed. Separate strategies are needed to work on these large volumes of data. Distributed processing of the data and combining the results (typically called **MapReduce**) is one strategy. It is also possible to process just enough data sequentially that can fit in a main memory each time and store the results somewhere on a hard drive; we need to repeat this process until the entirety of the data is processed completely. After the data processing, the results need to be combined to avail the final results of all the data that has been processed.

Special technologies such as Hadoop and Spark are required to perform ML on big data. Needless to say, you will need to hone specialized skills in order to apply ML algorithms successfully using these technologies on big data.

Natural language processing

This is an application area of ML that aims for computers to comprehend human languages such as English, French, and Mandarin. NLP applications enable users to interact with computers using spoken languages.

Chatbot, speech synthesis, machine translation, text classification and clustering, text generation, and text summarization are some of the popular applications of NLP.

Computer vision

This field of ML tries to mimic human vision. The aim is to enable computers to see, process, and determine the objects in images or videos. Deep learning and the availability of powerful hardware has led to the rise of very powerful applications in this area of ML.

Autonomous vehicles such as self-driving cars, object recognition, object tracking, motion analysis, and the restoration of images are some of the applications of computer vision.

Cost function

Cost function, loss function, or error function are used interchangeably by practitioners. Each is used to define and measure the error of a model. The objective for the ML algorithm is to minimize the loss from the dataset.

Some of the examples of cost function are square loss that is used in linear regression, hinge loss that is used in support vector machines and 0/1 loss used to measure accuracy in classification algorithms.

Model accuracy

Accuracy is one of the popular metrics used to measure the performance of ML models. The measurement is easy to understand and helps the practitioner to communicate the goodness of a model very easily to its business users.

Generally, this metric is used for classification problems. Accuracy is measured as the number of correct predictions divided by the total number of predictions.

Confusion matrix

This is a table that describes the classification model's performance. It is an n rows, n columns matrix where n represents the number of classes that are predicted by the classification model. It is formed by noting down the number of correct and incorrect predictions by the model when compared to the actual label.

Confusion matrices are better explained with an example—assume that there are 100 images in a dataset where there are 50 dog images and 50 cat images. A model that is built to classify images as cat images or dog images is given this dataset. The output from the model showed that 40 dog images are classified correctly and 20 cat images are predicted correctly. The following table is the confusion matrix construction from the prediction output of the model:

Model predicted labels	Actual labels		
		cats	dogs
	cats	20	30
	dogs	10	40

Predictor variables

These variables are otherwise called **independent variables** or **x-values**. These are the input variables that help to predict the dependent or target or response variable.

In a house rent prediction use case, the size of the house in square feet, the number of bedrooms, the number of houses available unoccupied in the region, the proximity to public transport, the accessibility to facilities such as hospitals and schools are all some examples of predictor variables that determine the rental cost of the house.

Response variable

Dependent variables or target or y-values are all interchangeably used by practitioners as alternatives for the term **response variable**. This is the variable the model predicts as output based on the independent variables that are provided as input to the model.

In the house rent prediction use case, the rent predicted is the response variable.

Dimensionality reduction

Feature reduction (or feature selection) or dimensionality reduction is the process of reducing the input set of independent variables to obtain a lesser number of variables that are really required by the model to predict the target.

In certain cases, it is possible to represent multiple dependent variables by combining them together without losing much information. For example, instead of having two independent variables such as the length of a rectangle and the breath of a rectangle, the dimensions can be represented by only one variable called the area that represents both the length and breadth of the rectangle.

The following mentioned are the multiple reasons we need to perform a dimensionality reduction on a given input dataset:

- To aid data compression, therefore accommodate the data in a smaller amount of disk space.
- The time to process the data is reduced as fewer dimensions are used to represent the data.
- It removes redundant features from datasets. Redundant features are typically known as **multicollinearity** in data.

- Reducing the data to fewer dimensions helps visualize the data through graphs and charts.
- Dimensionality reduction removes noisy features from the dataset which, in turn, improves the model performance.

There are many ways by which dimensionality reduction can be attained in a dataset. The use of filters, such as information gain filters, and symmetric attribute evaluation filters, is one way. Genetic-algorithm-based selection and **principal component analysis** (**PCA**) are other popular techniques used to achieve dimensionality reduction. Hybrid methods do exist to attain feature selection.

Class imbalance problem

Let's assume that one needs to build a classifier that identifies cat and dog images. The problem has two classes namely cat and dog. If one were to train a classification model, training data is required. The training data in this case is based on images of dogs and cats given as input so a supervised learning model can learn the features of dogs versus cats.

It may so happen that if there are 100 images available for training in the dataset and 95 of them are dog pictures, five of them are cat pictures. This kind of unequal representation of different classes in a training dataset is termed as a class imbalance problem.

Most ML techniques work best when the number of examples in each class are roughly equal. One can employ certain techniques to counter class imbalance problems in data. One technique is to reduce the majority class (images of dogs) samples and make them equal to the minority class (images of cats). In this case, there is information loss as a lot of the dog images go unused. Another option is to generate synthetic data similar to the data for the minority class (images of cats) so as to make the number of data samples equal to the majority class. **Synthetic minority over-sampling technique** (**SMOTE**) is a very popular technique for generating synthetic data.

It may be noted that accuracy is not a good metric for evaluating the performance of models where the training dataset experiences class imbalance problems. Assume a model built based on a class-imbalanced dataset that predicts a majority class for any test sample that it is asked to predict on. In this case, one gets 95% accuracy as roughly 95% of the images are dog images in the test dataset. But this performance can only be termed as a hoax as the model does not have any discriminative power—it just predicts dog as the class for any image it needs to predict about. In this case, it just happened that every image is predicted as a dog, but still the model got away with a very high accuracy indicating that it is a great model, whether it is in reality or not!

There are several other performance metrics available to use in a situation where a class imbalance is a problem, F1 score and the **area under the curve of the receiver operating characteristic (AUCROC)** are some of the popular ones.

Model bias and variance

While several ML algorithms are available to build models, model selection can be done on the basis of the bias and variance errors that the models produce.

Bias error occurs when the model has a limited capability to learn the true signals from a dataset provided as input to it. Having a highly biased model essentially means the model is consistent but inaccurate on average.

Variance errors occur when the models are too sensitive to the training datasets with which they are trained. Having high variance in a model essentially means that the trained model will produce high accuracies on any test dataset on average, but their predictions are inconsistent.

Underfitting and overfitting

Underfitting and overfitting are the concepts closely associated with bias and variance. These two are the biggest causes for the poor performance of the models, therefore a practitioner has to pay very close attention to these issues while building ML models.

A situation where the model does not perform well with both training data as well as test data is termed as underfitting. This situation can be detected by observing high training errors and test errors. Having an underfitting problem means that the ML algorithm chosen to fit the model is not suitable to model the features of the training data. Therefore, the only remedy is to try other kinds of ML algorithms to model the data.

Overfitting is a situation where the model learned the features of the training data so well that it fails to generalize on other unseen data. In an overfitting model, noise or random fluctuations in the training data are considered as true signals by the model and it looks for these patterns in unseen data as well, therefore impacting the poor model performance.

Overfitting is more prevalent in non-parametric and non-linear models such as decision trees, and neural networks. Pruning the trees is one remedy to overcome the problem. Another remedial measure is a technique called **dropout** where some of the features learned from the model are dropped randomly from the model therefore making the model more generalizable to unseen data. Regularization is yet another technique to resolve overfitting problems. This is attained by penalizing the coefficients of the model so that the model generalizes better. L1 penalty and L2 penalty are the types of penalties through which regularization can be performed in regression scenarios.

The goal for a practitioner is to ensure that the model neither overfits nor underfits. To achieve this, it is essential to learn when to stop training the ML data. One could plot the training error and validation error (an error that is measured on a small portion of the training dataset that is kept aside) on a chart and identify the point where the training data keeps decreasing, however the validation error starts to rise.

At times, obtaining performance measurement on training data and expecting a similar measurement to be obtained on unseen data may not work. A more realistic training and test performance estimate is to be obtained from a model by adopting a data-resampling technique called k-fold cross validation. The k in k-fold cross validation refers to a number; examples include 3-fold cross validation, 5-fold cross validation, and 10-fold cross validation. The **k-fold cross validation** technique involves dividing the training data into k parts and running the training process $k + 1$ times. In each iteration, the training is performed on $k - 1$ partitions of the data and the kth partition is used exclusively for testing. It may be noted that the kth partition for testing and $k - 1$ partitions for training are shuffled in each iteration, therefore the training data and testing data do not stay constant in each iteration. This approach enables getting a pessimistic measurement of performance that can be expected from the model on the unseen data in the future.

10-fold cross validation with 10 runs to obtain model performance is considered to be a gold standard estimate for a model's performance among practitioners. Estimating the model's performance in this way is always recommended in industrial setups and for critical ML applications.

Data preprocessing

This is essentially a step that is adopted in the early stages of an ML project pipeline. Data preprocessing involves transforming the raw data in a format that is acceptable as input by ML algorithms.

Feature hashing, missing values imputation, transforming variables from numeric to nominal, and vice versa, are a few data preprocessing steps among the numerous things that can be done to data during preprocessing.

Raw text documents' transformation into word vectors is an example of data preprocessing. The word vectors thus obtained can be fed to an ML algorithm to achieve documents classification or documents clustering.

Holdout sample

While working on a training dataset, a small portion of the data is kept aside for testing the performance of the models. The small portion of data is unseen data (not used in training), therefore one can rely on the measurements obtained for this data. The measurements obtained can be used to tune the parameters of the model or just to report out the performance of the model so as to set expectations in terms of what level of performance can be expected from the model.

It may be noted that the performance measurement reported out on the basis of a holdout sample is not as robust an estimate as that of a k-fold cross validation estimate. This is because there could be some unknown biases that could have crept in during the random split of the holdout set from the original dataset. Also, there are also no guarantees that the holdout dataset has a representation of all the classes involved in the training dataset. If we need representation of all classes in the holdout dataset, then a special technique called a **stratified holdout sample** needs to be applied. This ensures that there is representation for all classes in the holdout dataset. It is obvious that a performance measurement obtained from a stratified holdout sample is a better estimate of performance than that of the estimate of performance obtained from a nonstratified holdout sample.

70%-30%, 80%-20%, and 90%-10% are generally the sets of training data-holdout data splits observed in ML projects.

Hyperparameter tuning

ML or deep learning algorithms take hyperparameters as input prior to training the model. Each algorithm comes with its own set of hyperparameters and some algorithms may have zero hyperparameters.

Hyperparameter tuning is an important step in model building. Each of the ML algorithms comes with some default hyperparameter values that are generally used to build an initial model, unless the practitioner manually overrides the hyperparameters. Setting the right combination of hyperparameters and the right hyperparameter values for the model greatly improves the performance of the model in most cases. Hence, it is strongly recommended that one does hyperparameter tuning as part of ML model building. Searching through the possible universe of hyperparameter values is a very time-consuming task.

The *k* in k-means clustering and k-nearest neighbors classification, the number of tress and the depth of tress in random forest, and *eta* in XGBoost are all examples of hyperparameters.

Grid search and **Bayesian** optimization-based hyperparameter tuning are two popular methods of hyperparameter tuning among practitioners.

Performance metrics

A model needs to be evaluated on unseen data to assess its goodness. The term goodness may be expressed in several ways and these ways are termed as model performance metrics.

Several metrics exist to report the performance of models. Accuracy, precision, recall, F-score, sensitivity, specificity, AUROC curve, **root mean squared error** (**RMSE**), Hamming loss, and **mean squared error** (**MSE**) are some of the popular model performance metrics among others.

Feature engineering

Feature engineering is the art of creating new features either from existing data in the dataset or by procuring additional data from an external data source. It is done with the intent that adding additional features improves the model performance. Feature engineering generally requires domain expertise and in-depth business problem understanding.

Let's take a look at an example of feature engineering—for a bank that is working on a loan defaulter prediction project, sourcing and supplementing the training dataset with information on the unemployment trends of the region for the past few months might improve the performance of the model.

Model interpretability

Often, in a business environment when ML models are built, just reporting the performance measurements obtained to confirm the goodness of the model may not be enough. The stakeholders generally are inquisitive to understand the *whys* of the model, that is, what are the factors contributing to the model's performance? In other words, the stakeholders want to understand the causes of the effects. Essentially, the expectation from the stakeholders is to understand the importance of various features in the model and the direction in which each of the variables impacts the model.

For example, does a feature of *time spent on exercising every day* in the dataset for a cancer prediction model have any impact on the model predictions at all? If so, *does time spent on exercising every day* push the prediction in a negative direction or positive direction?

While the example might sound simple to generate an answer for, in real-world ML projects, model interpretability is not so very simple due to the complex relationships between variables. It is seldom that one feature, in its isolation, impacts the prediction in any one direction. It is indeed a **combination** of features that impact the prediction outcome. Thus, it is even more difficult to explain to what extent the feature is impacting the prediction.

Linear models are generally easier to explain even to business users. This is because we obtain weights for various features as a result of model training with linear algorithms. These weights are direct indicators of how a feature is contributing to model prediction. After all, in a linear model, a prediction is the linear combination of model weights and features passed through a function. It should be noted that interaction between variables in the real world are not essentially linear. So, a linear model trying to model the underlying data that has non-linear relationships may not have good predictive power. So, while linear models' interpretability is great, it comes at the cost of model performance.

On the contrary, non-linear and non-parametric models tend to be very difficult to interpret. In most cases, it may not be apparent even to the person building the models as to what exactly are the factors driving the prediction and in which direction. This is simply because the prediction outcome is a complex non-linear combination of variables. It is also known that non-linear models in general are better performing models when compared to linear models. Therefore, there is a trade-off needed between model interpretability and model performance.

While the goal of model interpretability is difficult to achieve, there is some merit in accomplishing this goal. It helps with the retrospection of a model that is deemed as being a good performing model and confirming that no noise inadvertently existed in the data that is used for model building and testing. It is obvious that models with noise as features fail to generalize on unseen data. Model interpretability helps with making sure that no noise crept into the models as features. Also, it helps build trust with business users that are eventually consumers of the model output. After all, there is no point in building a model whose output is not going to be consumed!

Non-parametric, non-linear models are difficult to interpret, if not impossible. Specialized ML methods are now available to aid black box models interpretability. **Partial dependency plot (PDP)**, **Locally interpretable model-agnostic explanations (LIME)**, and **Shapley additive explanations (SHAP)** also known as Sharpley's are some of the popular methods used by practitioners to decipher the internals of a black box model.

Now that there is a good understanding of the various fundamental terms of ML, our next journey is to explore the details of the ML project pipeline. This journey discussed in the next section helps us understand the process of building an ML project, deploying it, and obtaining predictions to use in a business.

ML project pipeline

Most of the content available on ML projects, either through books, blogs, or tutorials, explains the mechanics of machine learning in such a way that the dataset available is split into training, validation, and test datasets. Models are built using training datasets, and model improvements through hyperparameter tuning are done iteratively through validation data. Once a model is built and improved upon to a point that is acceptable, it is tested for goodness with unseen test data and the results of testing are reported out. Most of the public content available, ends at this point.

In reality, the ML projects in a business situation go beyond this step. We may observe that if one stops at testing and reporting a built model performance, there is no real use of the model in terms of predicting about data that is coming up in future. We also need to realize that the idea of building a model is to be able to deploy the model in production and have the predictions based on new data so that businesses can take appropriate action.

In a nutshell, the model needs to be saved and reused. This also means that any new data on which predictions need to be made needs to be preprocessed in the same way as training data. This ensures that, the new data has the same number of columns and also the same types of columns as training data. This part of productionalization of the models built in the lab is totally ignored when being taught. This section covers an end-to-end pipeline for the models, right from data preprocessing to building the models in the lab to productionalization of the models.

ML pipelines describe the entire process from raw data acquisition to obtaining post processing of the prediction results on unseen data so as to make it available for some kind of action by business. It is possible that a pipeline may be depicted at a generalized level or described at a very granular level. This current section focuses on describing a generic pipeline that may be applied to any ML project. Figure 1.8 shows the various components of the ML project pipeline otherwise known as the **cross-industry standard process for data mining (CRISP-DM)**.

Business understanding

Once the problem description that is to be solved using ML is clearly articulated, the first step in the ML pipeline is to be able to ascertain if the problem is of relevance to business and if the goals of the project are laid out without any ambiguities. It may also be wise to check if the problem at hand is feasible to be solved as an ML problem. These are the various aspects typically covered during the business understanding step.

Understanding and sourcing the data

The next step is to identify all the sources of data that are relevant to the business problem at hand. Organizations will have one or more systems, such as HR management systems, accounting systems, and inventory management systems. Depending on the nature of the problem at hand, we may need to fetch the data from multiple sources. Also, data that is obtained through the data acquisition step need not always be structured as tabular data; it could be unstructured data, such as emails, recorded voice files, and images.

In corporate organizations of reasonable size, it may not be possible for an ML professional to work all alone on the task of fetching the data from the diverse range of systems. Tight collaboration with other professionals in the organization may be required to complete this step of the pipeline successfully:

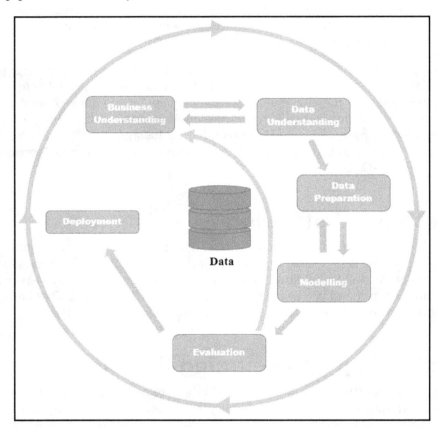

Preparing the data

Data preparation enables the creation of input data for ML algorithms to consume. Raw data that we get from data sources is often not very clean. Sometimes, the data cannot be readily fused into an ML algorithm to create a model. We need to ensure that the raw data is cleaned up and it is prepared in a format that is acceptable for the ML algorithm to take as input.

EDA is a substep in the process of creating the input data. It is a process of using visual and quantitative aids to understand the data without getting prejudice about the contents of the data. EDA gives us deeper insights into the data available at hand. It helps us to understand the required data preparation steps. Some of the insights that we could obtain during EDA are the existence of outliers in the data, missing values existence in the data, and the duplication of data. All of these problems are addressed during data cleansing which is another substep in data preparation. Several techniques may be adopted during data cleansing and the following mentioned are some of the popular techniques:

- Deleting records that are outliers
- Deleting redundant columns and irrelevant columns in data
- Missing values imputation—filling missing values with special value NA or a blank or median or mean or mode or with a regressed value
- Scaling the data
- Removing stop words such as *a, and*, and *how*, from unstructured text data
- Normalizing words in unstructured text documents with techniques such as stemming, and lemmatization
- Eliminating non-dictionary words in text data
- Spelling corrections on misspelled words in text documents
- Replacing non-recognizable domain-specific acronyms in the text with actual word descriptions
- Rotation, scaling, and translation of image data

Representing the unstructured data as vectors, providing labels for the records if the problem at hand needs to be dealt with by supervised learning, handling class imbalance problems in the data, feature engineering, transforming the data through transformation functions such as log transform, min-max transform, square root transform, and cube transform, are all part of the data preparation process.

The output of the data preparation step is tabular data that can be fit readily into an ML algorithm as input in order to create models.

An additional substep that is typically done in data preparation is to divide the dataset into training data, validation data, and test data. These various datasets are used for specific purposes in the model-building step.

Model building and evaluation

Once the data is ready and prior to the creation of the model, we need to pick and select the features from the list of features available. This can be accomplished through several off-the-shelf feature selection techniques. Some ML algorithms (for example, XGBoost) have feature selection inbuilt within the algorithm, therefore we need not explicitly perform feature selection prior to carrying out the modeling activity.

A suite of ML algorithms is available to try and create models on the data. Additionally, models may be created through ensembling techniques as well. One needs to pick the algorithm(s) and create models using training datasets, then tune the hyperparameters of the model using validation datasets. Finally, the model that is created can be tested using the test dataset. Issues, such as selecting the right metric to evaluate model performance, overfitting, underfitting, and acceptable performance thresholds all need to be taken care of in the model-building step itself. It may be noted that if we do not obtain acceptable performance on the model, it is required to go back to previous steps in order to get more data or to create additional features and then repeat the model-building step once again to check if the model performance improves. This may be done as many times as required until the desired level of performance is achieved by the model.

At the end of the modeling step, we might end up with a suite of models each having its own performance measurement on the unseen test data. The model that has the best performance can be selected for use in production.

Model deployment

The next step is to save the final model that can be used for the future. There are several ways to save the model as an object. Once saved, the model can be reloaded any time and it can be used for scoring the new data. Saving the model as an object is a trivial task and a number of libraries are available in Python and R to achieve it. As a result of saving the model, the model object gets persisted to the disk as a `.sav` file or a `.pkl` file or a `.pmml` object depending on the library used. The object can then be loaded into the memory to perform scoring on unseen data.

The final model that is selected for use in production can be deployed to score unseen data in the following two modes:

- **Batch mode**: Batch mode scoring is when one accumulates the unseen data to be scored in a file, then run a batch job (just another executable script) at a predetermined time to perform scoring. The job loads the model object from disk to the memory and runs on each of the records in the file that needs to be scored. The output is written to another file at a specified location as directed in the batch job script. It may be noted that the records to be scored should have the same number of columns as in the training data and the type of columns should also comply with the training data. It should be ensured that the number of levels in factors columns (nominal type data) should also match with that of the training data.

- **Real-time mode**: There are times where the business needs model scoring to happen on the fly. In this case, unlike the batch mode, data is not accumulated and we do not wait until the batch job runs for scoring. The expectation is that each record of the data, as and when it is available for scoring should be scored by the model. The result of the scoring is to be available to business users almost instantaneously. In this case, a model needs to be deployed as a web service that can serve any requests that come in. The record to be scored can be passed to the web service through a simple API call which, in turn, returns the scored result that can be consumed by the downstream applications. Again, the unscored data record that is passed in the API call should comply with the format of the training data records.

Yet another way of achieving near real-time results is by running the model job on micro batches of data several times a day and at very frequent intervals. The data gets accumulated between the intervals until a point where the model job kicks off. The model job scores and outputs the results for the data that is accumulated similar to batch mode. The business user gets to see the scored results as soon as the micro batch job finishes execution. The only difference between the micro batches processing versus the batch is that unlike the batch mode, business users need not wait until the next business day to get the scored results.

Though, the model building pipeline ends with successfully deploying the ML model and making it available for scoring, in real-world business situations, the job does not end here. Of course, the success parties flow in but there is a need to look again at the models post a certain point in time (maybe in a few months post the deployment). A model that is not maintained at regular intervals does not get very well used by businesses.

To avoid the models from perishing and not being used by business users, it is important to collect feedback on the performance of the model over a period of time and capture if any improvements need to be incorporated in the models. The unseen data does not come with labels, therefore comparing the model output with that of the desired output by business is a manual exercise. Collaborating with business users is a strong requirement to get feedback in this situation.

If there is a continued business need for the model and if the performance is not up to the mark on the unseen data that is scored with existing model, it needs to be investigated to identify the root cause(s). It may so happen that several things have changed in the data that is scored over a period of time when compared to the data on which model was initially trained. In which case, there is a strong need to recalibrate the model and it is essentially a jolly good idea to start once again!

Now that the book has covered all the essentials of ML and the project pipeline, the next topic to be covered is the learning paradigm, which will help us learn several ML algorithms.

Learning paradigm

Most learning paradigms that are followed in other books or content about ML follow a bottom-up approach. This approach starts from the bottom and works its way up. The approach first covers the theoretical elements, such as mathematical introductions to the algorithm, the evolution of the algorithm, variations, and parameters that the algorithm takes, and then delves into the application of the ML algorithm specific to a dataset. This may be a good approach; however, it takes longer really to see the results produced by the algorithm. It needs a lot of perseverance on the part of the learner to be patient and wait until the practical application of the algorithm is covered. In most cases, practitioners and certain classes of industry professionals working on ML are really interested in the practical aspects and they want to experience the power of the algorithm. For these people, the focus is not the theoretical foundations of the algorithm, but it is the practical application. The bottom-up approach works counterproductively in this case.

The learning paradigm followed in this book to teach several ML algorithms is opposite to the bottom-up approach. It rather follows a very practical top-down approach. The focus of this approach is **learning by coding**.

Each chapter of the book will focus on learning a particular class of ML algorithm. To start with, the chapter focuses on how to use the algorithm in various situations and how to obtain results from the algorithm in practice. Once the practical application of the algorithm is demonstrated using code and a dataset, gradually, the rest of the chapter unveils the theoretical details/concepts of the algorithms experienced in the chapter thus far. All theoretical details will be ensured to be covered only in as much detail as is required to understand the code and to apply the algorithm on any new datasets. This ensures that we get to learn the focused application areas of the algorithms rather than unwanted theoretical aspects that are of less importance applied in the ML world.

Datasets

Each chapter of the book describes an ML project that solves a business problem using an ML algorithm or a set of algorithms that we attempt to learn in that specific chapter. The projects considered are from different domains ranging from health care, to banking and finance, and to robots. The business problems solved in the chapters that follow are carefully selected to demonstrate solving a close-to-real-world business use case. The datasets used for the problems are popular open datasets. This will help us not only to explore the solutions covered in this book but also to examine other solutions that are developed for the problem. The problem solved in each of the chapters enriches our experience by applying ML algorithms in various domains and helps us get an understanding of how to solve the business problems in various domains successfully.

Summary

Well! We have learned so much together so far, and now we have reached the end of this chapter. In this chapter, we covered all that deals with ML, including the terminologies and the project pipeline. We also talked about the learning paradigm, the datasets, and all the topics and projects that will be covered in each chapter.

In the next chapter, we will start to work on ML ensembles to predict employee attrition.

2
Predicting Employee Attrition Using Ensemble Models

If you reviewed the recent machine learning competitions, one key observation I am sure you would make is that the recipes of all three winning entries in most of the competitions include very good feature engineering, along with well-tuned ensemble models. One conclusion I derive from this observation is that good feature engineering and building well-performing models are two areas that should be given equal emphasis in order to deliver successful machine learning solutions.

While feature engineering most times is something that is dependent on the creativity and domain expertise of the person building the model, building a well-performing model is something that can be achieved through a philosophy called **ensembling**. Machine learning practitioners often use ensembling techniques to beat the performance benchmarks yielded by even the best performing individual ML algorithm. In this chapter, we will learn about the following topics of this exciting area of ML:

- Philosophy behind ensembling
- Understanding the attrition problem and the dataset
- K-nearest neighbors model for benchmarking the performance
- Bagging
- Randomization with random forests
- Boosting
- Stacking

Philosophy behind ensembling

Ensembling, which is super-famous among ML practitioners, can be well-understood through a simple real-world, non-ML example.

Assume that you have applied for a job in a very reputable corporate organization and you have been called for an interview. It is unlikely you will be selected for a job just based on one interview with an interviewer. In most cases, you will go through multiple rounds of interviews with several interviewers or with a panel of interviewers. The expectation from the organization is that each of the interviewers is an expert on a particular area and that the interviewer has evaluated your fitness for the job based on your experience in the interviewers' area of expertise. Your selection for the job, of course, depends on consolidated feedback from all of the interviewers that talked to you. The organization deems that you will be more successful in the job as your selection is based on a consolidated decision made by multiple experts and not just based on one expert's decision, which may be prone to certain biases.

Now, when we talk about the consolidation of feedback from all the interviewers, the consolidation can happen through several methods:

- **Averaging**: Assume that your candidature for the job is based on you clearing a cut-off score in the interviews. Assume that you have met ten interviewers and each one of them have rated you on a maximum score of 10 which represents your experience as perceived by interviewers in his area of expertise. Now, your consolidated score is made by simply averaging all your scores given by all the interviewers.

- **Majority vote**: In this case, there is no actual score out of 10 which is provided by each of the interviewers. However, of the 10 interviewers, eight of them confirmed that you are a good fit for the position. Two interviewers said no to your candidature. You are selected for the job as the majority of the interviewers are happy with your interview performance.

- **Weighted average**: Let's consider that four of the interviewers are experts in some minor skills that are good to have for the job you applied for. These are not mandatory skills needed for the position. You are interviewed by all 10 interviewers and each one of them have given you a score out of 10. Similar to the averaging method, in the weighted averaging method as well, your interviews final score is obtained by averaging the scores given by all interviewers.

However, not all scores are treated equally to compute the final score. Each interview score is multiplied with a weight and a product is obtained. All the products thus obtained thereby are summed to obtain the final score. The weight for each interview is a function of the importance of the skill it tested in the candidate and the importance of that skill to do the job. It is obvious that a *good to have* skill for the job carries a lower weight when compared to a *must have* skill. The final score now inherently represents the proportion of mandatory skills that the candidate possesses and this has more influence on your selection.

Similar to the interviews analogy, ensembling in ML also produces models based on consolidated learning. The term **consolidated learning** essentially represents learning obtained through applying several ML algorithms or it is learning obtained from several data subsets that are part of a large dataset. Analogous to interviews, multiple models are learned from the application of ensembling technique. However, a final consolidation is arrived at regarding the prediction by means of applying one of the averaging, majority voting, or weighted averaging techniques on individual predictions made by each of the individual models. The models created from the application of an ensembling technique along with the prediction consolidation technique is typically termed as an **ensemble**.

Each ML algorithm is special and has a unique way to model the underlying training data. For example, a k-nearest neighbors algorithm learns by computing distances between the elements in dataset; naive Bayes learns by computing the probabilities of each attribute in the data belonging to a particular class. Multiple models may be created using different ML algorithms and predictions can be done by combining predictions of several ML algorithms. Similarly, when a dataset is partitioned to create subsets and if multiple models are trained using an algorithm each focusing on one dataset, each model is very focused and it is specialized in learning the characteristics of the subset of data it is trained on. In both cases, with models based on multiple algorithms and multiple subsets of data, when we combine the predictions of multiple models through consolidation, we get better predictions as we leverage multiple strengths that each model in an ensemble carry. This, otherwise, is not obtained when using a single model for predictions.

The crux of ensembling is that, better predictions are obtained when we combine the predictions of multiple models than just relying on one model for prediction. This is no different from the management philosophy that together we do better, which is otherwise termed as **synergy**!

Now that we understand the core philosophy behind ensembling, we are now ready to explore the different types of ensembling techniques. However, we will learn the ensembling techniques by implementing them in a project to predict the attrition of employees. As we already know, prior to building any ML project, it is very important to have a deep understanding of the problem and the data. Therefore, in the next section, we first focus on understanding the attrition problem at hand, then we study the dataset associated with the problem, and lastly we understand the properties of the dataset through exploratory data analysis (EDA). The key insights we obtain in this section come from a one-time exercise and will hold good for all the ensembling techniques we will apply in the later sections.

Getting started

To get started with this section, you will have to download the `WA_Fn-UseC_-HR-Employee-Attrition.csv` dataset from the GitHub link for the code in this chapter.

Understanding the attrition problem and the dataset

HR analytics helps with interpreting organizational data. It finds out the people-related trends in the data and helps the HR department take the appropriate steps to keep the organization running smoothly and profitably. Attrition in a corporate setup is one of the complex challenges that the people managers and HR personnel have to deal with. Interestingly, machine learning models can be deployed to predict potential attrition cases, thereby helping the appropriate HR personnel or people managers take the necessary steps to retain the employee.

In this chapter, we are going to build ML ensembles that will predict such potential cases of attrition. The job attrition dataset used for the project is a fictional dataset created by data scientists at IBM. The `rsample` library incorporates this dataset and we can make use of this dataset directly from the library.

It is a small dataset that has 1,470 records of 31 attributes. The description of the dataset can be obtained with the following code:

```
setwd("~/Desktop/chapter 2")
library(rsample)
data(attrition)
str(attrition)
mydata<-attrition
```

This will result in the following output:

```
'data.frame':1470 obs. of  31 variables:
 $ Age                     : int   41 49 37 33 27 32 59 30 38 36 ...
 $ Attrition               : Factor w/ 2 levels "No","Yes": 2 1 2 1 1 1 1 1
1 1 ....
 $ BusinessTravel          : Factor w/ 3 levels "Non-
Travel","Travel_Frequently",..: 3 2 3 2 3 2 3 3 2 3 ...
 $ DailyRate               : int   1102 279 1373 1392 591 1005 1324 1358 216
1299 ...
 $ Department              : Factor w/ 3 levels "Human_Resources",..: 3 2 2
2 2 2 2 2 2 2 ...
 $ DistanceFromHome        : int   1 8 2 3 2 2 3 24 23 27 ...
 $ Education               : Ord.factor w/ 5 levels "Below_College"<..: 2 1
2 4 1 2 3 1 3 3 ...
 $ EducationField          : Factor w/ 6 levels "Human_Resources",..: 2 2 5
2 4 2 4 2 2 4 ...
 $ EnvironmentSatisfaction : Ord.factor w/ 4 levels "Low"<"Medium"<..: 2 3
4 4 1 4 3 4 4 3 ...
 $ Gender                  : Factor w/ 2 levels "Female","Male": 1 2 2 1 2
2 1 2 2 2 ...
 $ HourlyRate              : int   94 61 92 56 40 79 81 67 44 94 ...
 $ JobInvolvement          : Ord.factor w/ 4 levels "Low"<"Medium"<..: 3 2
2 3 3 3 4 3 2 3 ...
 $ JobLevel                : int   2 2 1 1 1 1 1 1 3 2 ...
 $ JobRole                 : Factor w/ 9 levels
"Healthcare_Representative",..: 8 7 3 7 3 3 3 3 5 1 ...
 $ JobSatisfaction         : Ord.factor w/ 4 levels "Low"<"Medium"<..: 4 2
3 3 2 4 1 3 3 3 ...
 $ MaritalStatus           : Factor w/ 3 levels "Divorced","Married",..: 3
2 3 2 2 3 2 1 3 2 ...
 $ MonthlyIncome           : int   5993 5130 2090 2909 3468 3068 2670 2693
9526 5237 ...
 $ MonthlyRate             : int   19479 24907 2396 23159 16632 11864 9964
13335 8787 16577 ...
 $ NumCompaniesWorked      : int   8 1 6 1 9 0 4 1 0 6 ...
 $ OverTime                : Factor w/ 2 levels "No","Yes": 2 1 2 2 1 1 2 1
1 1 ...
 $ PercentSalaryHike       : int   11 23 15 11 12 13 20 22 21 13 ...
```

```
 $ PerformanceRating       : Ord.factor w/ 4 levels
"Low"<"Good"<"Excellent"<..: 3 4 3 3 3 3 4 4 4 3 ...
 $ RelationshipSatisfaction: Ord.factor w/ 4 levels "Low"<"Medium"<..: 1 4
2 3 4 3 1 2 2 2 ...
 $ StockOptionLevel        : int  0 1 0 0 1 0 3 1 0 2 ...
 $ TotalWorkingYears       : int  8 10 7 8 6 8 12 1 10 17 ...
 $ TrainingTimesLastYear   : int  0 3 3 3 2 3 2 2 3 ...
 $ WorkLifeBalance         : Ord.factor w/ 4 levels
"Bad"<"Good"<"Better"<..: 1 3 3 3 3 2 2 3 3 2 ...
 $ YearsAtCompany          : int  6 10 0 8 2 7 1 1 9 7 ...
 $ YearsInCurrentRole      : int  4 7 0 7 2 7 0 0 7 7 ...
 $ YearsSinceLastPromotion : int  0 1 0 3 2 3 0 0 1 7 ...
 $ YearsWithCurrManager    : int  5 7 0 0 2 6 0 0 8 7 ...
```

To view the `Attrition` target variable in the dataset run the following code:

```
table(mydata$Attrition)
```

This will result in the following output:

```
  No   Yes
1233   237
```

Out of the 1,470 observations in the dataset, we have 1,233 samples (83.87%) that are non-attrition cases and 237 attrition cases (16.12%). Clearly, we are dealing with a *class imbalance* dataset.

We will now visualize the highly correlated variables in the data through the `corrplot` library using the following code:

```
# considering only the numeric variables in the dataset
numeric_mydata <-
mydata[,c(1,4,6,7,10,11,13,14,15,17,19,20,21,24,25,26,28:35)]
# converting the target variable "yes" or "no" values into numeric
# it defaults to 1 and 2 however converting it into 0 and 1 to be
consistent
numeric_Attrition = as.numeric(mydata$Attrition)- 1
# create a new data frame with numeric columns and numeric target
numeric_mydata = cbind(numeric_mydata, numeric_Attrition)
# loading the required library
library(corrplot)
# creating correlation plot
M <- cor(numeric_mydata)
corrplot(M, method="circle")
```

This will result in the following output:

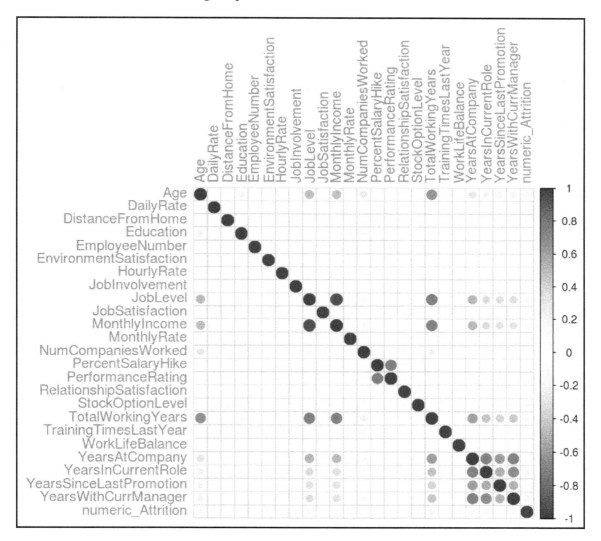

In the preceding screenshot, it may be observed that darker and larger blues dot in the cells indicate the existence of a strong correlation between the variables in the corresponding rows and columns that form the cell. High correlation between the independent variables indicates the existence of redundant features in the data. The problem of the existence of highly correlated features in the data is termed as **multicollinearity**. If we were to fit a regression model, then it is required that we treat the highly correlated variables from the data through some techniques such as removing the redundant features or by applying principal component analysis or partial least squares regression, which intuitively cuts down the redundant features.

We infer from the output that the following variables are highly correlated and the person building the model needs to take care of these variables if we are to build a regression-based model:

```
JobLevel-MonthlyIncome; JobLevel-TotalWorkingYears; MonthlyIncome-
TotalWorkingYears; PercentSalaryHike-PerformanceRating; YearsAtCompany-
YearsInCurrentRole; YearsAtCompany-
YearsWithCurrManager; YearsWithCurrManager-YearsInCurrentRole
```

Now, plot the various independent variables with the dependent `Attrition` variable in order to understand the influence of the independent variable on the target:

```
### Overtime vs Attiriton
l <- ggplot(mydata, aes(OverTime, fill = Attrition))
l <- l + geom_histogram(stat="count")

tapply(as.numeric(mydata$Attrition) - 1 ,mydata$OverTime,mean)

No Yes
0.104364326375712 0.305288461538462
```

Let's run the following command to get a graph view:

```
print(l)
```

The preceding command generates the following output:

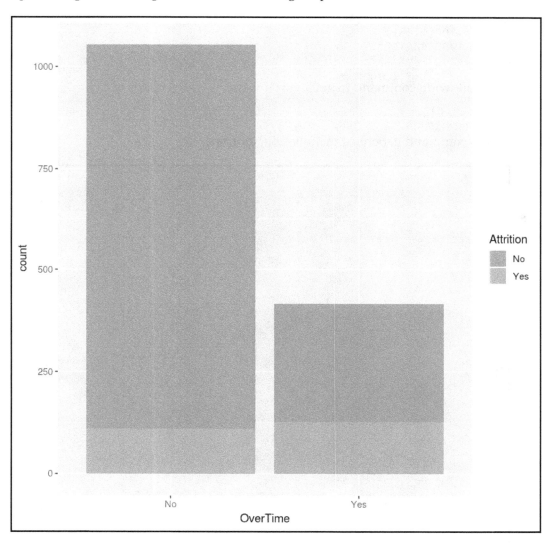

In the preceding output, it can be observed that employees that work overtime are more prone to attrition when compared to the ones that do not work overtime:

Let's calculate the attrition of the employees by executing the following commands:

```
### MaritalStatus vs Attiriton
l <- ggplot(mydata, aes(MaritalStatus,fill = Attrition))
```

```
l <- l + geom_histogram(stat="count")

tapply(as.numeric(mydata$Attrition) - 1 ,mydata$MaritalStatus,mean)
Divorced 0.100917431192661
Married 0.12481426448737
Single 0.25531914893617
```

Let's run the following command to get a graph view:

```
print(l)
```

The preceding command generates the following output:

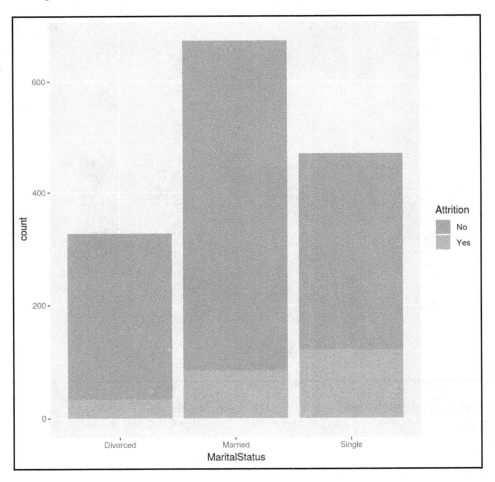

In the preceding output, it can be observed that employees that are single have more attrition:

```
###JobRole vs Attrition
l <- ggplot(mydata, aes(JobRole,fill = Attrition))
l <- l + geom_histogram(stat="count")

tapply(as.numeric(mydata$Attrition) - 1 ,mydata$JobRole,mean)

Healthcare Representative      Human Resources
              0.06870229      0.23076923
     Laboratory Technician     Manager
              0.23938224      0.04901961
     Manufacturing Director    Research Director
              0.06896552      0.02500000
        Research Scientist     Sales Executive
              0.16095890      0.17484663
       Sales Representative
              0.39759036
mean(as.numeric(mydata$Attrition) - 1)
[1] 0.161224489795918
```

Execute the following command to get a graphical representation for the same:

```
print(l)
```

Take a look at the following output generated by running the preceding command:

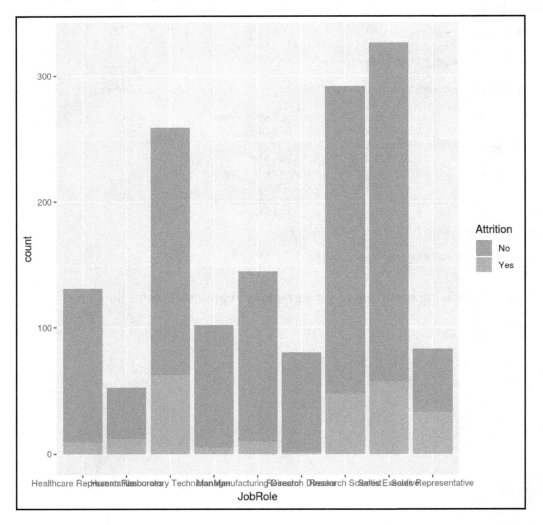

In the preceding output, it can be observed that the lab technicians, sales representatives, and employees working in human resources job roles have more attrition than other organizational roles.

Let's execute the following commands to check with the impact of the gender of an employee over attribution:

```
###Gender vs Attrition
l <- ggplot(mydata, aes(Gender,fill = Attrition))
```

```
l <- l + geom_histogram(stat="count")

tapply(as.numeric(mydata$Attrition) - 1 ,mydata$Gender,mean)

Female 0.147959183673469
Male 0.170068027210884
```

Run the following command to get a graphical representation for the same:

```
print(l)
```

This will result in the following output:

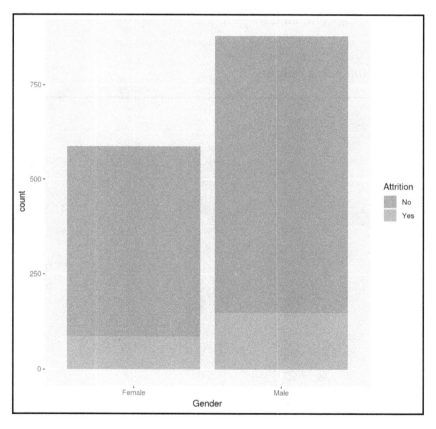

In the preceding output, you can see that the gender of an employee does not have any impact on attrition, in other words attrition is observed to be the same across all genders.

Let's calculate the attribute of the employees from various fields by executing the following:

```
###EducationField vs Attrition el <- ggplot(mydata, aes(EducationField,fill
= Attrition))
l <- l + geom_histogram(stat="count")

tapply(as.numeric(mydata$Attrition) - 1 ,mydata$EducationField,mean)

Human Resources    Life Sciences    Marketing
      0.2592593       0.1468647      0.2201258
        Medical    Other Technical  Degree
      0.1357759       0.1341463      0.2424242
```

Let's execute the following command to get a graphical representation:

```
print(l)
```

This will result in the following output:

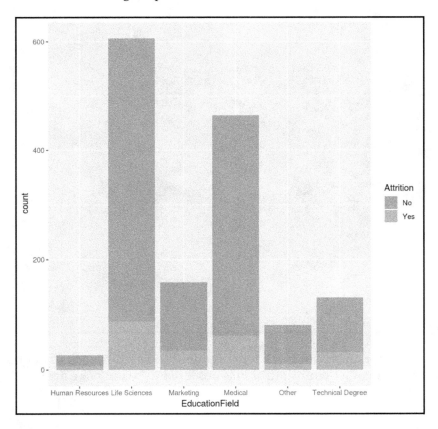

Looking at the preceding graph, we can conclude that employees with a technical degree or a degree in human resources are observed to have more attrition. Take a look at the following code:

```
###Department vs Attrition
l <- ggplot(mydata, aes(Department,fill = Attrition))
l <- l + geom_histogram(stat="count")

tapply(as.numeric(mydata$Attrition) - 1 ,mydata$Department,mean)
Human Resources   Research & Development  Sales
    0.1904762          0.1383975              0.2062780
```

Let's execute the following command to check with the attribution of various departments:

```
print(l)
```

This will result in the following output:

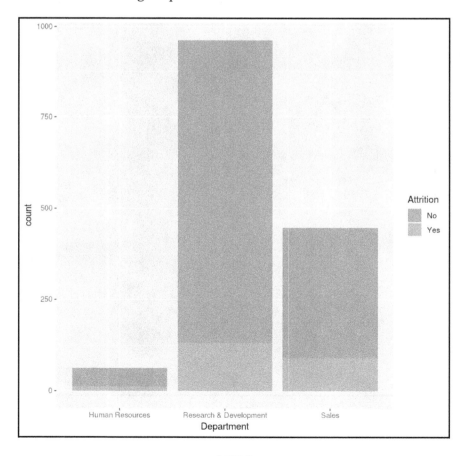

Looking at the preceding graph, we can conclude that the R and D department has less attrition compared to the sales and HR departments. Take a look at the following code:

```
###BusinessTravel vs Attrition
l <- ggplot(mydata, aes(BusinessTravel,fill = Attrition))
l <- l + geom_histogram(stat="count")

tapply(as.numeric(mydata$Attrition) - 1 ,mydata$BusinessTravel,mean)
 Non-Travel    Travel_Frequently    Travel_Rarely
  0.0800000    0.2490975              0.1495686
```

Execute the following command to get a graphical representation for the same:

```
print(l)
```

This will result in the following output:

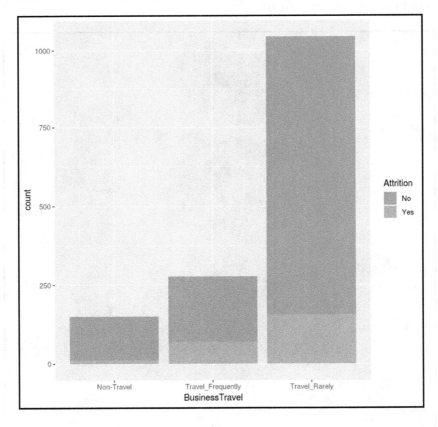

Looking at the preceding graph, we can conclude that employees with frequent travels are prone to more attrition compared to employees with a non-travel status or the ones that rarely travel.

Let's calculate the overtime of the employees by executing the following commands:

```
### x=Overtime, y= Age, z = MaritalStatus , t = Attrition
ggplot(mydata, aes(OverTime, Age)) +
  facet_grid(.~MaritalStatus) +
  geom_jitter(aes(color = Attrition),alpha = 0.4) +
  ggtitle("x=Overtime, y= Age, z = MaritalStatus , t = Attrition") +
  theme_light()
```

This will result in the following output:

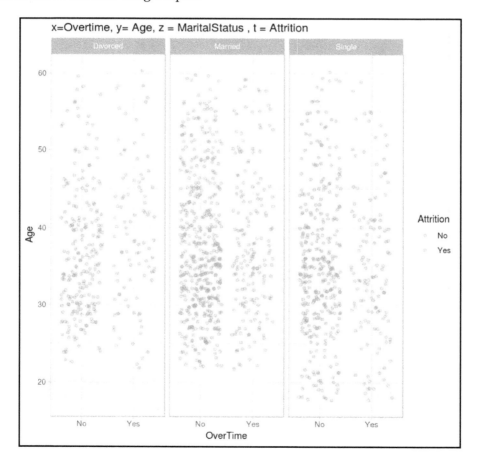

Looking at the preceding graph, we can conclude that it can be observed that employees that are young (age < 35) and are single, but work overtime, are more prone to attrition:

```
### MonthlyIncome vs. Age, by  color = Attrition
ggplot(mydata, aes(MonthlyIncome, Age, color = Attrition)) +
  geom_jitter() +
  ggtitle("MonthlyIncome vs. Age, by  color = Attrition ") +
  theme_light()
```

This will result in the following output:

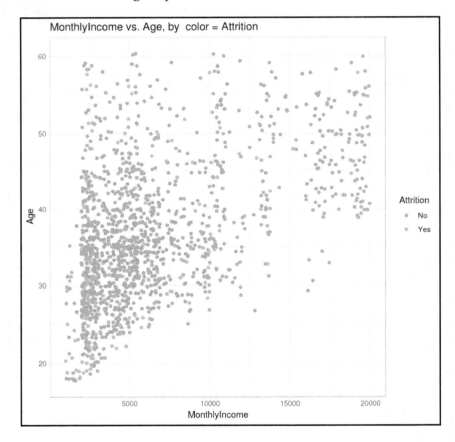

Looking at the preceding graph, we can conclude that attrition is higher in employees that are young (age < 30) and most attrition is observed with employees that earn less than $7,500.

Although we have learned a number of important details about the data at hand, there is actually so much more to explore and learn. However, so as to move to the next step, we stop here at this EDA step. It should be noted that, in a real-world situation, data would not be so very clean as we see in this attrition dataset. For example, we would have missing values in the data; in which case, we would do missing values imputation. Fortunately, we have an impeccable dataset that is ready for us to create models without having to do any data cleansing or additional preprocessing.

K-nearest neighbors model for benchmarking the performance

In this section, we will implement the **k-nearest neighbors** (**KNN**) algorithm to build a model on our IBM attrition dataset. Of course, we are already aware from EDA that we have a class imbalance problem in the dataset at hand. However, we will not be treating the dataset for class imbalance for now as this is an entire area on its own and several techniques are available in this area and therefore out of scope for the ML ensembling topic covered in this chapter. We will, for now, consider the dataset as is and build ML models. Also, for class imbalance datasets, Kappa or precision and recall or the area under the curve of the receiver operating characteristic (AUROC) are the appropriate metrics to use. However, for simplicity, we will use *accuracy* as a performance metric. We will adapt 10-fold cross validation repeated 10 times to avail the model performance measurement. Let's now build our attrition prediction model with the KNN algorithm as follows:

```
# Load the necessary libraries
# doMC is a library that enables R to use multiple cores available on the
sysem thereby supporting multiprocessing.
library(doMC)
# registerDoMC command instructs R to use the specified number of cores to
execute the code. In this case, we ask R to use 4 cores available on the
system
registerDoMC(cores=4)
# caret library has the ml algorithms and other routines such as cross
validation etc.
library(caret)
# Setting the working directory where the dataset is located
setwd("~/Desktop/chapter 2")
# Reading the csv file into R variable called mydata
mydata <- read.csv("WA_Fn-UseC_-HR-Employee-Attrition.csv")
#Removing the non-discriminatory features (as identified during EDA) from
the dataset
mydata$EmployeeNumber=mydata$Over18=mydata$EmployeeCount=mydata$StandardHou
rs = NULL
```

```
# setting the seed prior to model building ensures reproducibility of the
results obtained
set.seed(10000)
# setting the train control parameters specifying gold standard 10 fold
cross validation  repeated 10 times
fitControl = trainControl(method="repeatedcv", number=10,repeats=10)
###creating a model on the data. Observe that we specified Attrition as the
target and that model should learn from rest of the variables. We specified
mydata as the dataset to learn. We pass the train control parameters and
specify that knn algorithm need to be used to build the model. K can be of
any length - we specified 20 as parameter which means the train command
will search through 20 different random k values and finally retains the
model that produces the best performance measurements. The final model is
stored as caretmodel
caretmodel = train(Attrition~., data=mydata, trControl=fitControl, method =
"knn", tuneLength = 20)
# We output the model object to the console
caretmodel
```

This will result in the following output:

```
k-Nearest Neighbors
1470 samples
  30 predictors
   2 classes: 'No', 'Yes'
No pre-processing
Resampling: Cross-Validated (10 fold, repeated 10 times)
Summary of sample sizes: 1323, 1323, 1324, 1323, 1324, 1322, ...
Resampling results across tuning parameters:
  k    Accuracy   Kappa
   5   0.8216447  0.0902934591
   7   0.8349033  0.0929511324
   9   0.8374198  0.0752842114
  11   0.8410920  0.0687849122
  13   0.8406861  0.0459679081
  15   0.8406875  0.0337742424
  17   0.8400748  0.0315670261
  19   0.8402770  0.0245499585
  21   0.8398721  0.0143638854
  23   0.8393945  0.0084393721
  25   0.8391891  0.0063246624
  27   0.8389174  0.0013913143
  29   0.8388503  0.0007113939
  31   0.8387818  0.0000000000
  33   0.8387818  0.0000000000
  35   0.8387818  0.0000000000
  37   0.8387818  0.0000000000
  39   0.8387818  0.0000000000
```

```
41   0.8387818   0.0000000000
43   0.8387818   0.0000000000
Accuracy was used to select the optimal model using the largest value.
The final value used for the model was k = 11.
```

We can see from the model output that the best performing model is when k = 11 and we obtained an accuracy of 84% with this k value. In the rest of the chapter, while experimenting with several ensembling techniques, we will check if this 84% accuracy obtained from KNN will get beaten at all.

In a realistic project-building situation, just identifying the best hyperparameters is not enough. A model needs to be trained on a full dataset with the best hyperparameters and the model needs to be saved for future use. We will review these steps in the rest of this section.

In this case, the caretmodel object already has the trained model with k = 11, therefore we do not attempt to retrain the model with the best hyperparameter. To check the final model, you can query the model object with the code:

```
caretmodel$finalModel
```

This will result in the following output:

```
11-nearest neighbor model
Training set outcome distribution:
  No   Yes
1233   237
```

The next step is to save your best models to a file so that we can load them up later and make predictions on unseen data. A model can be saved to a local directory using the saveRDS R command:

```
# save the model to disk
saveRDS(caretmodel, "production_model.rds")
```

In this case, the caretmodel is saved as production_model.rds in the working directory. The model is now serialized as a file that can be loaded anytime and it can be used to score unseen data. Loading and scoring can be achieved through the following R code:

```
# Set the working directory to the directory where the saved .rds file is
located
setwd("~/Desktop/chapter 2")
#Load the model
loaded_model <- readRDS("production_model.rds")
```

```
#Using the loaded model to make predictions on unseen data
final_predictions <- predict(loaded_model, unseen_data)
```

 Please note that `unseen_data` needs to be read prior to scoring through the `predict` command.

The part of the code where the final model is trained on the entire dataset, saving the model, reloading it from the file whenever required and scoring the unseen data collectively, is termed as building an ML productionalization pipeline. This pipeline remains the same for all ML models irrespective of the fact that the model is built using one single algorithm or using an ensembling technique. Therefore, in the later sections when we implement the various ensembling techniques, we will not cover the productionalization pipeline but just stop at obtaining the performance measurement through 10-fold cross validation repeated 10 times.

Bagging

Bootstrap aggregation or **bagging** is the earliest ensemble technique adopted widely by the ML-practicing community. Bagging involves creating multiple different models from a single dataset. It is important to understand an important statistical technique called bootstrapping in order to get an understanding of bagging.

Bootstrapping involves multiple random subsets of a dataset being created. It is possible that the same data sample gets picked up in multiple subsets and this is termed as **bootstrapping with replacement**. The advantage with this approach is that the standard error in estimating a quantity that occurs due to the use of whole dataset. This technique can be better explained with an example.

Assume you have a small dataset of 1,000 samples. Based on the samples, you are asked to compute the average of the population that the sample represents. Now, a direct way of doing it is through the following formula:

$$Average = sum(all\ 1000\ samples)/1000$$

As this is a small sample, we may have an error in estimating the population average. This error can be reduced by adapting bootstrap sampling with replacement. In the technique, we create 10 subsets of the dataset where each dataset has 100 items in it. A data item may be randomly represented multiple times in a subset and there is no restriction on the number of times an item can be represented within a data subset as well as across the subsets. Now, we take the average of samples in each data subset, therefore, we end up with 10 different averages. Using all these collected averages, we estimate the average of the population with the following formula:

$$Average = sum(averages\ obtained\ from\ each\ of\ the\ 10\ sub\ datasets)/10$$

Now, we have a better estimate of the average as we have extrapolated the small sample to randomly generate multiple samples that are representative of the original population.

In bagging, the actual training dataset is split into multiple bags through bootstrap sampling with replacement. Assuming that we ended up with n bags, when an ML algorithm is applied on each of these bags, we obtain n different models. Each model is focused on one bag. When it comes to making predictions on new unseen data, each of these n models makes independent predictions on the data. A final prediction for an observation is arrived at by combining the predictions of the observation of all the n models. In case of classification, voting is adopted and the majority is considered as the final prediction. For regression, the average of predictions from all models is considered as the final prediction.

Decision-tree-based algorithms, such as **classification and regression trees** (**CART**), are unstable learners. The reason is that a small change in the training dataset heavily impacts the model created. Model change essentially means that the predictions also change. Bagging is a very effective technique to handle the high sensitivity to data changes. As we can build multiple decision tree models on subsets of a dataset and then arrive at a final prediction based on predictions from each of the models, the effect of changes in data gets nullified or not experienced very significantly.

One intuitive problem experienced with building multiple models on subsets of data is **overfitting**. However, this is overcome by growing deep trees without applying any pruning on the nodes.

A downside with bagging is that it takes longer to build the models when compared to building a model with a stand-alone ML algorithm. This is obvious because multiple models gets built in bagging, as opposed to one single model, and it takes time to build these multiple models.

Now, let's implement the R code to achieve a bagging ensemble and compare the performance obtained with that of the performance obtained from KNN. We will then explore the working mechanics of bagging methodology.

The `caret` library provides a framework to implement bagging with any stand-alone ML algorithm. `ldaBag`, `plsBag`, `nbBag`, `treeBag`, `ctreeBag`, `svmBag`, and `nnetBag` are some of the example methods provided in caret. In this section, we will implement bagging with three different `caret` methods such as `treebag`, `svmbag`, and `nbbag`.

Bagged classification and regression trees (treeBag) implementation

To begin, load the essential libraries and register the number of cores for parallel processing:

```
library(doMC)
registerDoMC(cores = 4)
library(caret)
#setting the random seed for replication
set.seed(1234)
# setting the working directory where the data is located
setwd("~/Desktop/chapter 2")
# reading the data
mydata <- read.csv("WA_Fn-UseC_-HR-Employee-Attrition.csv")
#removing the non-discriminatory features identified during EDA
mydata$EmployeeNumber=mydata$Over18=mydata$EmployeeCount=mydata$StandardHou
rs = NULL
#setting up cross-validation
cvcontrol <- trainControl(method="repeatedcv", repeats=10, number = 10,
allowParallel=TRUE)
# model creation with treebag , observe that the number of bags is set as
10
train.bagg <- train(Attrition ~ ., data=mydata, method="treebag",B=10,
trControl=cvcontrol, importance=TRUE)
train.bagg
```

This will result in the following output:

```
Bagged CART
1470 samples
  30 predictors
   2 classes: 'No', 'Yes'
No pre-processing
Resampling: Cross-Validated (10 fold, repeated 10 times)
```

```
Summary of sample sizes: 1324, 1323, 1323, 1322, 1323, 1322, ...
Resampling results:
  Accuracy  Kappa
  0.854478  0.2971994
```

We can see that we achieved a better accuracy of 85.4% compared to 84% accuracy that was obtained with the KNN algorithm.

Support vector machine bagging (SVMBag) implementation

The steps of loading the libraries, registering multiprocessing, setting a working directory, reading data from a working directory, removing nondiscriminatory features from data, and setting up cross-validation parameters remain the same in the SVMBag and NBBag implementations as well. So, we do not repeat these steps in the SVMBag or NBBag code. Rather, we will focus on discussing the SVMBag or NBBag specific code:

```
# Setting up SVM predict function as the default svmBag$pred function has
some code issue
svm.predict <- function (object, x)
{
 if (is.character(lev(object))) {
    out <- predict(object, as.matrix(x), type = "probabilities")
    colnames(out) <- lev(object)
    rownames(out) <- NULL
  }
  else out <- predict(object, as.matrix(x))[, 1]
  out
}
# setting up parameters to build svm bagging model
bagctrl <- bagControl(fit = svmBag$fit,
                      predict = svm.predict ,
                      aggregate = svmBag$aggregate)
# fit the bagged svm model
set.seed(300)
svmbag <- train(Attrition ~ ., data = mydata, method="bag",trControl =
cvcontrol, bagControl = bagctrl,allowParallel = TRUE)
# printing the model results
svmbag
```

This will result in the following output:

```
Bagged Model

1470 samples
  30 predictors
   2 classes: 'No', 'Yes'

No pre-processing
Resampling: Cross-Validated (10 fold, repeated 10 times)
Summary of sample sizes: 1324, 1324, 1323, 1323, 1323, 1323, ...
Resampling results:
  Accuracy   Kappa
  0.8777721  0.4749657

Tuning parameter 'vars' was held constant at a value of 44
```

You will see that we achieved an accuracy of 87.7%, which is much higher than the KNN model's 84% accuracy.

Naive Bayes (nbBag) bagging implementation

We will now do the nbBag implementation by executing the following code:

```
# setting up parameters to build svm bagging model
bagctrl <- bagControl(fit = nbBag$fit,
                      predict = nbBag$pred ,
                      aggregate = nbBag$aggregate)
# fit the bagged nb model
set.seed(300)
nbbag <- train(Attrition ~ ., data = mydata, method="bag", trControl =
cvcontrol, bagControl = bagctrl)
# printing the model results
nbbag
```

This will result in the following output:

```
Bagged Model

1470 samples
  30 predictors
   2 classes: 'No', 'Yes'

No pre-processing
Resampling: Cross-Validated (10 fold, repeated 10 times)
Summary of sample sizes: 1324, 1324, 1323, 1323, 1323, 1323, ...
```

```
Resampling results:

  Accuracy   Kappa
  0.8389878  0.00206872

Tuning parameter 'vars' was held constant at a value of 44
```

We see that in this case, we achieved only 83.89% accuracy, which is slightly inferior to the KNN model's performance of 84%.

Although we have shown only three examples of the `caret` methods for bagging, the code remains the same to implement the other methods. The only change that is needed in the code is to replace the `fit`, `predict`, and `aggregate` parameters in `bagControl`. For example, to implement bagging with a neural network algorithm, we need to define `bagControl` as follows:

```
bagControl(fit = nnetBag$fit, predict = nnetBag$pred , aggregate =
nnetBag$aggregate)
```

It may be noted that an appropriate library needs to be available in R for `caret` to run the methods, otherwise it results in error. For example, `nbBag` requires the `klaR` library to be installed on the system prior to executing the code. Similarly, the `ctreebag` function needs the `party` package to be installed. Users need to check the availability of an appropriate library on the system prior to including it for use with the `caret` bagging.

We now have an understanding of implementing a project through bagging technique. The next subsection covers the underlying working mechanism of bagging. This will help get clarity in terms of what bagging did internally with our dataset so as to produce better performance measurements than that of stand-alone model performance.

Randomization with random forests

As we've seen in bagging, we create a number of bags on which each model is trained. Each of the bags consists of subsets of the actual dataset, however the number of features or variables remain the same in each of the bags. In other words, what we performed in bagging is subsetting the dataset rows.

In random forests, while we create bags from the dataset through subsetting the rows, we also subset the features (columns) that need to be included in each of the bags.

Assume that you have 1,000 observations with 20 features in your dataset. We can create 20 bags where each one of the bags has 100 observations (this is possible because of bootstrapping with replacement) and five features. Now 20 models are trained where each model gets to see only the bag it is assigned with. The final prediction is arrived at by voting or averaging based on the fact of whether the problem is a regression problem or a classification problem.

Another key difference between bagging and random forests is the ML algorithm that is used to build the model. In bagging, any ML algorithm may be used to create a model however random forest models are built specifically using CART.

Random forest modeling is yet another very popular machine learning algorithm. It is one of the algorithms that has proved itself multiple times as the best performing of algorithms, despite applying it on noisy datasets. For a person that has understood bootstrapping, understanding random forests is a cakewalk.

Implementing an attrition prediction model with random forests

Let's get our attrition model through random forest modeling by executing the following code:

```
# loading required libraries and registering multiple cores to enable
parallel processing
library(doMC)
library(caret)
registerDoMC(cores=4)
# setting the working directory and reading the dataset
setwd("~/Desktop/chapter 2")
mydata <- read.csv("WA_Fn-UseC_-HR-Employee-Attrition.csv")
# removing the non-discriminatory features from the dataset as identified
during EDA step
mydata$EmployeeNumber=mydata$Over18=mydata$EmployeeCount=mydata$StandardHou
rs = NULL
# setting the seed for reproducibility
set.seed(10000)
# setting the cross validation parameters
fitControl = trainControl(method="repeatedcv", number=10,repeats=10)
# creating the caret model with random forest algorithm
caretmodel = train(Attrition~., data=mydata, method="rf",
trControl=fitControl, verbose=F)
# printing the model summary
caretmodel
```

This will result in the following output:

```
Random Forest

1470 samples
  30 predictors
   2 classes: 'No', 'Yes'

No pre-processing
Resampling: Cross-Validated (10 fold, repeated 10 times)
Summary of sample sizes: 1323, 1323, 1324, 1323, 1324, 1322, ...
Resampling results across tuning parameters:

  mtry  Accuracy   Kappa
   2    0.8485765  0.1014859
  23    0.8608271  0.2876406
  44    0.8572929  0.2923997

Accuracy was used to select the optimal model using the largest value.
The final value used for the model was mtry = 23.
```

We see the best random forest model achieved a better accuracy of 86% compared to KNN's 84%.

Boosting

A weak learner is an algorithm that performs relatively poorly—generally, the accuracy obtained with the weak learners is just above chance. It is often, if not always, observed that weak learners are computationally simple. Decision stumps or 1R algorithms are some examples of weak learners. Boosting converts weak learners into strong learners. This essentially means that boosting is not an algorithm that does the predictions, but it works with an underlying weak ML algorithm to get better performance.

A boosting model is a sequence of models learned on subsets of data similar to that of the bagging ensembling technique. The difference is in the creation of the subsets of data. Unlike bagging, all the subsets of data used for model training are not created prior to the start of the training. Rather, boosting builds a first model with an ML algorithm that does predictions on the entire dataset. Now, there are some misclassified instances that are subsets and used by the second model. The second model only learns from this misclassified set of data curated from the first model's output.

The second model's misclassified instances become input to the third model. The process of building models is repeated until the stopping criteria is met. The final prediction for an observation in the unseen dataset is arrived by averaging or voting the predictions from all the models for that specific, unseen observation.

There are subtle differences between the various and numerous algorithms in the boosting algorithms family, however we are not going to discuss them in detail as the intent of this chapter is to get a generalized understanding of ML ensembles and not to gain in-depth knowledge of various boosting algorithms.

While obtaining better performance, measurement is the biggest advantage with the boosting ensemble; difficulty with model interpretability, higher computational times, and model overfitting are some of the issues encountered with boosting. Of course, these problems can be overruled through the use of specialized techniques.

Boosting algorithms are undoubtedly super-popular and are observed to be used by winners in many Kaggle and similar competitions. There are a number of boosting algorithms available such as **gradient boosting machines** (**GBMs**), **adaptive boosting** (**AdaBoost**) , gradient tree boosting, **extreme gradient boosting** (**XGBoost**), and **light gradient boosting machine** (**LightGBM**). In this section, we will learn the theory and implementation of two of the most popular boosting algorithms such as GBMs and XGBoost. Prior to learning the theoretical concept of boosting and its pros and cons, let's first start focusing on implementing the attrition prediction models with GBMs and XGBoost.

The GBM implementation

Let's implement the attrition prediction model with GBMs:

```
# loading the essential libraries and registering the cores for
multiprocessing
library(doMC)
library(mlbench)
library(gbm)
library(caret)
registerDoMC(cores=4)
# setting the working directory and reading the dataset
setwd("~/Desktop/chapter 2")
mydata <- read.csv("WA_Fn-UseC_-HR-Employee-Attrition.csv")
# removing the non-discriminatory features as identified by EDA step
mydata$EmployeeNumber=mydata$Over18=mydata$EmployeeCount=mydata$StandardHou
rs = NULL
# converting the target attrition feild to numeric as gbm model expects all
```

```
numeric feilds in the dataset
mydata$Attrition = as.numeric(mydata$Attrition)
# forcing the attrition column values to be 0 and 1 instead of 1 and 2
mydata = transform(mydata, Attrition=Attrition-1)
# running the gbm model with 10 fold cross validation to identify the
number of trees to build - hyper parameter tuning
gbm.model = gbm(Attrition~., data=mydata, shrinkage=0.01, distribution =
'bernoulli', cv.folds=10, n.trees=3000, verbose=F)
# identifying and printing the value of hyper parameter identified through
the tuning above
best.iter = gbm.perf(gbm.model, method="cv")
print(best.iter)
# setting the seed for reproducibility
set.seed(123)
# creating a copy of the dataset
mydata1=mydata
# converting target to a factor
mydata1$Attrition=as.factor(mydata1$Attrition)
# setting up cross validation controls
fitControl = trainControl(method="repeatedcv", number=10,repeats=10)
# runing the gbm model in tandem with caret
caretmodel = train(Attrition~., data=mydata1, method="gbm",
distribution="bernoulli",  trControl=fitControl, verbose=F,
tuneGrid=data.frame(.n.trees=best.iter, .shrinkage=0.01,
.interaction.depth=1, .n.minobsinnode=1))
# printing the model summary
print(caretmodel)
```

This will result in the following output:

```
2623
Stochastic Gradient Boosting

1470 samples
  30 predictors
   2 classes: '0', '1'

No pre-processing
Resampling: Cross-Validated (10 fold, repeated 10 times)
Summary of sample sizes: 1323, 1323, 1323, 1322, 1323, 1323, ...
Resampling results:
  Accuracy   Kappa
  0.8771472  0.4094991
Tuning parameter 'n.trees' was held constant at a value of 2623
Tuning parameter 'shrinkage' was held constant at a value of 0.01
Tuning parameter 'n.minobsinnode' was held constant at a value of 1
```

You will see that with the GBM model, we have achieved accuracy above 87%, which is better accuracy compared to the 84% achieved with KNN.

Building attrition prediction model with XGBoost

Now, let's implement the attrition prediction model with XGBoost:

```
# loading the required libraries and registering the cores for
multiprocessing
library(doMC)
library(xgboost)
library(caret)
registerDoMC(cores=4)
# setting the working directory and loading the dataset
setwd("~/Desktop/chapter 2")
mydata <- read.csv("WA_Fn-UseC_-HR-Employee-Attrition.csv")
# removing the non-discriminatory features from the dataset as identified
in EDA step
mydata$EmployeeNumber=mydata$Over18=mydata$EmployeeCount=mydata$StandardHou
rs = NULL
# setting up cross validation parameters
ControlParamteres <- trainControl(method = "repeatedcv",number = 10,
repeats=10, savePredictions = TRUE, classProbs = TRUE)
# setting up hyper parameters grid to tune
parametersGrid <-  expand.grid(eta = 0.1, colsample_bytree=c(0.5,0.7),
max_depth=c(3,6),nrounds=100, gamma=1, min_child_weight=2,subsample=0.5)
# printing the parameters grid to get an intuition
print(parametersGrid)
# xgboost model building
modelxgboost <- train(Attrition~., data = mydata, method = "xgbTree",
trControl = ControlParamteres, tuneGrid=parametersGrid)
# printing the model summary
print(modelxgboost)
```

This will result in the following output:

eta	colsample_bytree	max_depth	nrounds	gamma	min_child_weight	subsample
0.1	0.5	3	100	1	2	0.5
0.1	0.7	3	100	1	2	0.5
0.1	0.5	6	100	1	2	0.5
0.1	0.7	6	100	1	2	0.5

```
eXtreme Gradient Boosting
1470 samples
```

```
  30 predictors
   2 classes: 'No', 'Yes'

No pre-processing
Resampling: Cross-Validated (10 fold, repeated 10 times)
Summary of sample sizes: 1323, 1323, 1322, 1323, 1323, 1322, ...
Resampling results across tuning parameters:

  max_depth  colsample_bytree  Accuracy   Kappa
  3          0.5               0.8737458  0.3802840
  3          0.7               0.8734728  0.3845053
  6          0.5               0.8730674  0.3840938
  6          0.7               0.8732589  0.3920721

Tuning parameter 'nrounds' was held constant at a value of 100
Tuning parameter 'min_child_weight' was held constant at a value of 2
Tuning parameter 'subsample' was held constant at a value of 0.5
Accuracy was used to select the optimal model using the largest value.
The final values used for the model were nrounds = 100, max_depth = 3, eta
= 0.1, gamma = 1, colsample_bytree = 0.5, min_child_weight = 2 and
subsample = 0.5.
```

Again, we observed that with XGBoost model, we have achieved an accuracy above 87%, which is a better accuracy compared to the 84% achieved with KNN.

Stacking

In all the ensembles we have learned about so far, we have manipulated the dataset in certain ways and exposed subsets of the data for model building. However, in stacking, we are not going to do anything with the dataset; instead we are going to apply a different technique that involves using multiple ML algorithms instead. In stacking, we build multiple models with various ML algorithms. Each algorithm possesses a unique way of learning the characteristics of data and the final stacked model indirectly incorporates all those unique ways of learning. Stacking gets the combined power of several ML algorithms through getting the final prediction by means of voting or averaging as we do in other types of ensembles.

Building attrition prediction model with stacking

Let's build an attrition prediction model with stacking:

```
# loading the required libraries and registering the cpu cores for
multiprocessing
library(doMC)
library(caret)
library(caretEnsemble)
registerDoMC(cores=4)
# setting the working directory and loading the dataset
setwd("~/Desktop/chapter 2")
mydata <- read.csv("WA_Fn-UseC_-HR-Employee-Attrition.csv")
# removing the non-discriminatory features from the dataset as identified
in EDA step
mydata$EmployeeNumber=mydata$Over18=mydata$EmployeeCount=mydata$StandardHou
rs = NULL
# setting up control paramaters for cross validation
control <- trainControl(method="repeatedcv", number=10, repeats=10,
savePredictions=TRUE, classProbs=TRUE)
# declaring the ML algorithms to use in stacking
algorithmList <- c('C5.0', 'nb', 'glm', 'knn', 'svmRadial')
# setting the seed to ensure reproducibility of the results
set.seed(10000)
# creating the stacking model
models <- caretList(Attrition~., data=mydata, trControl=control,
methodList=algorithmList)
# obtaining the stacking model results and printing them
results <- resamples(models)
summary(results)
```

This will result in the following output:

```
summary.resamples(object = results)

Models: C5.0, nb, glm, knn, svmRadial
Number of resamples: 100

Accuracy
                 Min.    1st Qu.     Median       Mean     3rd Qu.       Max. NA's
C5.0       0.8082192  0.8493151  0.8639456  0.8625833  0.8775510  0.9054054     0
nb         0.8367347  0.8367347  0.8378378  0.8387821  0.8424658  0.8435374     0
glm        0.8299320  0.8639456  0.8775510  0.8790444  0.8911565  0.9387755     0
knn        0.8027211  0.8299320  0.8367347  0.8370763  0.8438017  0.8630137     0
svmRadial  0.8287671  0.8648649  0.8775510  0.8790467  0.8911565  0.9319728     0

Kappa  Min.         1st Qu.     Median       Mean     3rd Qu.       Max. NA's
C5.0   0.03992485  0.29828006  0.37227344  0.3678459  0.4495049  0.6112590     0
```

```
nb        0.00000000  0.00000000  0.00000000  0.0000000  0.0000000  0.0000000      0
glm       0.26690604  0.39925723  0.47859218  0.4673756  0.5218094  0.7455280      0
knn      -0.05965697  0.02599388  0.06782465  0.0756081  0.1320451  0.2431312      0
svmRadial 0.24565     0.38667527  0.44195662  0.4497538  0.5192393  0.7423764      0
```

```
# Identifying the correlation between results
modelCor(results)
```

This will result in the following output:

	C5.0	nb	glm	knn	svmRadial
C5.0	1.00000000	0.06912034	0.4728593	0.19511949	0.45963498
nb	0.06912034	1.00000000	0.1128155	0.07580389	0.06687541
glm	0.47285929	0.11281554	1.0000000	0.15578044	0.53965278
knn	0.19511949	0.07580389	0.1557804	1.00000000	0.23502484
svmRadial	0.45963498	0.06687541	0.5396528	0.23502484	1.00000000

We can see from the correlation table results that none of the individual ML algorithm predictions are highly correlated. Very highly correlated results mean that the algorithms have produced very similar predictions. Combining the very similar predictions may not really yield significant benefit compared with what one would avail from accepting the individual predictions. In this specific case, we can observe that none of the algorithm predictions are highly correlated so we can straightforwardly move to the next step of stacking the predictions:

```
# Setting up the cross validation control parameters for stacking the
predictions from individual ML algorithms
stackControl <- trainControl(method="repeatedcv", number=10, repeats=10,
savePredictions=TRUE, classProbs=TRUE)
# stacking the predictions of individual ML algorithms using generalized
linear model
stack.glm <- caretStack(models, method="glm", trControl=stackControl)
# printing the stacked final results
print(stack.glm)
```

This will result in the following output:

```
A glm ensemble of 2 base models: C5.0, nb, glm, knn, svmRadial
Ensemble results:
Generalized Linear Model
14700 samples
    5 predictors
    2 classes: 'No', 'Yes'
No pre-processing
```

```
Resampling: Cross-Validated (10 fold, repeated 10 times)
Summary of sample sizes: 13230, 13230, 13230, 13230, 13230, 13230, ...
Resampling results:
  Accuracy   Kappa
  0.8844966  0.4869556
```

With GLM-based stacking, we have 88% accuracy. Let's now examine the effect of using random forest modeling instead of GLM to stack the individual predictions from each of the five ML algorithms on the observations:

```
# stacking the predictions of individual ML algorithms using random forest
stack.rf <- caretStack(models, method="rf", trControl=stackControl)
# printing the summary of rf based stacking
print(stack.rf)
```

This will result in the following output:

```
A rf ensemble of 2 base models: C5.0, nb, glm, knn, svmRadial
Ensemble results:
Random Forest
14700 samples
    5 predictors
    2 classes: 'No', 'Yes'
No pre-processing
Resampling: Cross-Validated (10 fold, repeated 10 times)
Summary of sample sizes: 13230, 13230, 13230, 13230, 13230, 13230, ...
Resampling results across tuning parameters:
  mtry  Accuracy   Kappa
  2     0.9122041  0.6268108
  3     0.9133605  0.6334885
  5     0.9132925  0.6342740
Accuracy was used to select the optimal model using the largest value.
The final value used for the model was mtry = 3.
```

We see that without much effort, we were able to achieve an accuracy of 91% by stacking the predictions. Now, let's explore the working principle of stacking.

At last, we have discovered the various ensembling techniques that can provide us with better performing models. However, before ending the chapter, there are a couple of things we need to take a note of.

There is not just one way to implement ML models in R. For example, bagging can be implemented using functions available in the `ipred` library and not by using `caret` as we did in this chapter. We should be aware that hyperparameter tuning forms an important part of model building to avail the best performing model. The number of hyperparameters and the acceptable values for those hyperparameters vary depending on the library that we intend to use. This is the reason why we paid less attention to hyperparameter tuning in the models we built in this chapter. Nevertheless, it is very important to read up the library documentation to understand the hyperparameters that can be tuned with a library function. In most cases, incorporating hyperparameter tuning in models significantly improves the model's performance.

Summary

To recollect, we were using a class-imbalanced dataset to build the attrition model. Using techniques to resolve the class imbalance prior to model building is another key aspect of getting better model performance measurements. We used bagging, randomization, boosting, and stacking to implement and predict the attrition model. We were able to accomplish 91% accuracy just by using the features that were readily available in the models. Feature engineering is a crucial aspect whose role cannot be ignored in ML models. This may be one other path to explore to improve model performance further.

In the next chapter, we will explore the secret recipe of recommending products or content through building a personalized recommendation engines. I am all set to implement a project to recommend jokes. Turn to the next chapter to continue the journey of learning.

3
Implementing a Jokes Recommendation Engine

I am sure this is something you have experienced as well: while shopping for a cellphone on Amazon, you are also shown some product recommendations of mobile accessories, such as screen guards and phone cases. Not very surprisingly, most of us end up buying one or more of these recommendations! The primary purpose of a recommendation engine in an e-commerce site is to lure buyers into purchasing more from vendors. Of course, this is no different from a salesperson trying to up-sell or cross-sell to customers in a physical store.

You may recollect the **Customers Who Bought This Item Also Bought This** heading on Amazon (or any e-commerce site) where recommendations are shown. The aim of these recommendations is to get you to buy not just one product but a product combo, therefore pushing the sales revenues in an upward direction. Recommendations on Amazon are so successful that McKinsey estimated that a whopping 35% of the overall sales made on Amazon is due to their recommendations!

In this chapter, we will learn about the theory and implementation of a recommendation engine to suggest jokes to users. To do this, we use the Jester's jokes dataset that is available in the `recommenderlab` library of R. We will cover the following major topics:

- Fundamental aspects of recommendation engines
- Understanding the Jokes recommendation problem and the dataset
- Recommendation system using an item-based collaborative filtering technique
- Recommendation system using a user-based collaborative filtering technique
- Recommendation system using an association-rule mining technique
- Content-based recommendation engine
- Hybrid recommendation system for Jokes recommendation

Fundamental aspects of recommendation engines

While the basic intent of showing recommendations is to push sales, they actually serve just beyond the better sales concept. Highly personalized content is something recommendation engines are able to deliver. This essentially means that recommendation engines on a retail platform such as Amazon are able to offer the right content to the right customer at the right time through the right channel. It makes sense to provide personalized content; after all, there is no point in showing an irrelevant product to a customer. Also, with the lower attention spans of customers, businesses want to be able to maximize their selling opportunities by showing the right products and encouraging them to buy the right products. At a very high level, personalized content recommendation is achieved in AI in several ways:

- **Mapping similar products that were bought together**: Let's take an example of an online shopper who searched for school bags on a shopping website. Very likely, the shopper would be interested in buying additional school-related items when buying a school bag. Therefore, displaying school bags along with notebooks, pencils, pens, and pencil cases ensures a higher probability of additional sales.

- **Recommendations based on customer demographics**: Showing high-end phones and stylish phone accessories as recommended products to conservative middle class customers, who generally look for steal deals, may not fetch a big upswing in sales of the recommended products. Instead, such customers might find these irrelevant recommendations to be annoying, therefore impacting their loyalty.

- **Recommendations based on similarities between customers**: Product recommendations to a customer are based on the products purchased or liked by other, similar customers. For example, recommending a newly-arrived cosmetic product to young women living in urban locations. The recommendation in this case is not just because of the attributes of the customer but because other customers of a similar type have bought this product. As the item grows in popularity among similar individuals, the product is chosen as the one to be recommended.

- **Recommendations based on product similarities**: If you search for a laptop backpack of a particular brand, along with the results of the searched item, you are also shown other brand laptop backpacks as recommendations. This recommendation is purely based on the similarity between the products.

- **Recommendations based on the historical purchase profile of customers**: If a customer has always purchased a particular brand of jeans, they are shown recommendations of newer varieties of jeans of the particular brand they tend to purchase. These recommendations are purely based on the historical purchases of the customer.
- **Hybrid recommendations**: It is possible that one or more recommendation approaches can be combined to arrive at the best recommendations for a customer. For example, a recommendation list can be arrived by using customer preferences inferred from the historical data as well as from the demographics information of the customer.

Repurchase campaigns, newsletter recommendations, rebinding the sales from abandoned carts, customized discounts and offers, and smoothened browsing experience of e-commerce sites are some of the applications of recommendation systems in the online retail industry.

Due to several prevalent use cases, it might appear that recommender systems are used in only in the e-commerce industry. However, this is not true. The following are some of the use cases of recommender systems in non e-commerce domains:

- In the pharmaceutical industry, recommender systems are applied to identify drugs patients with certain characteristics that they will respond better to
- Stocks recommendation are done based on the stock picks of a successful group of people
- YouTube and online media use a recommendation engine to serve content that is similar to the content currently being watched by the user
- Tourism recommendations are based on tourist spots that the user or similar users have visited
- Identifying skills and personality traits of future employees in various roles
- In the culinary sciences, dishes that go pair together can be explored through the application of recommender systems

The list can grow to an enormous size, given that use cases for recommendation systems exist in almost every domain.

Now that we have a basic understanding of the concept of recommendation systems and the value it offers to business, we can now move to our next section, where we attempt to understand the Jester's Jokes recommendation dataset and the problems that could be solved by building a recommendation engine.

Recommendation engine categories

Prior to implementing our first recommender system, let's explore the types of recommender systems in detail. The following diagram shows the broad categories of recommender systems:

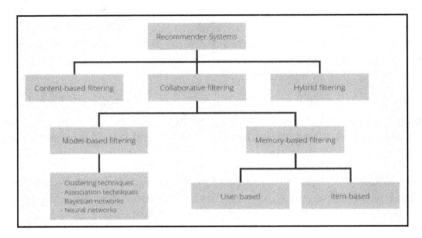

Recommender system categories

Each of the techniques shown in the diagram may be used to build a recommender system model that is capable of suggesting jokes to the users. Let's briefly explore the various recommendation engine categories.

Content-based filtering

Cognitive filtering, or content-based filtering, recommends items by comparing product attributes and customer profile attributes. The attributes of each product is represented as a set of tags or terms—typically the words that occur in a product description document. The customer profile is represented with the same terms and built by analyzing the content of products that have been seen or rated by the customer.

Collaborative filtering

Social filtering, or collaborative filtering, filters information by using the recommendations of other people. The principle behind collaborative filtering is that the customers who have appreciated the same items in the past have a high probability of displaying similar interests in the future as well.

We generally ask for reviews and recommendation from friends prior to watching a movie. A recommendation from a friend is more accepted than recommendations from others as we share some interests with our friends. This is the same principle on which collaborative filtering works.

Collaborative filtering can be further classified into memory-based and model-based as follows:

- **Memory-based**: In this method, user rating information is used to compute the likeness between users or items. This computed likeness is then used to come up with recommendations.
- **Model based**: Data mining methods are applied to recognize patterns in the data, and the learned patterns are then used to generate recommendations.

Hybrid filtering

In this class of recommendation systems, we combine more than one type of recommendation system to come up with final recommendations.

Getting started

To get started, you will have to download the supporting files from the GitHub link.

Understanding the Jokes recommendation problem and the dataset

Dr. Ken Goldberg and his colleagues, Theresa Roeder, Dhruv Gupta, and Chris Perkins, introduced a dataset to the world through their paper *Eigentaste: A Constant Time Collaborative Filtering Algorithm*, which is pretty popular in the recommender-systems domain. The dataset is named the Jester's jokes dataset. To create it, a number of users are presented with several jokes and they are asked to rate them. The ratings provided by the users for the various jokes formed the dataset. The data in this dataset is collected between April 1999 and May 2003. The following are the attributes of the dataset:

- Over 11,000,000 ratings of 150 jokes from 79,681 users
- Each row is a user (Row 1 = User #1)

- Each column is a joke (Column 1 = Joke #1)
- Ratings are given as real values from -10.00 to +10.00; -10 being the lowest possible rating and 10 being the highest
- 99 corresponds to a null rating

The `recommenderlab` package in R provides a subset of this original dataset provided by Dr. Ken Goldberg's group. We will make use of this subset for our projects covered in this chapter.

The `Jester5k` dataset provided in the `recommenderlab` library contains a 5,000 x 100 rating matrix (5,000 users and 100 jokes) with ratings between -10.00 and +10.00. All selected users have rated 36 or more jokes. The dataset is in the `realRatingMatrix` format. This is a special matrix format that the `recommenderlab` expects the data to be in, to apply the various functions that are packaged in the library.

As we are already aware, **exploratory data analysis (EDA)** is the first step for any data science project. Going by this principle, let's begin by reading the data, and then proceed with the EDA step on the dataset:

```
# including the required libraries
library(data.table)
library(recommenderlab)
# setting the seed so as to reproduce the results
set.seed(54)
# reading the data to a variable
library(recommenderlab)
data(Jester5k)
str(Jester5k)
```

This will result in the following output:

```
Formal class 'realRatingMatrix' [package "recommenderlab"] with 2 slots
  ..@ data       :Formal class 'dgCMatrix' [package "Matrix"] with 6 slots
  .. .. ..@ i       : int [1:362106] 0 1 2 3 4 5 6 7 8 9 ...
  .. .. ..@ p       : int [1:101] 0 3314 6962 10300 13442 18440 22513 27512
32512 35685 ...
  .. .. ..@ Dim     : int [1:2] 5000 100
  .. .. ..@ Dimnames:List of 2
  .. .. .. ..$ : chr [1:5000] "u2841" "u15547" "u15221" "u15573" ...
  .. .. .. ..$ : chr [1:100] "j1" "j2" "j3" "j4" ...
  .. .. ..@ x       : num [1:362106] 7.91 -3.2 -1.7 -7.38 0.1 0.83 2.91
-2.77 -3.35 -1.99 ...
  .. .. ..@ factors : list()
  ..@ normalize: NULL
```

The data structure output is pretty self explanatory and we see it provides empirical evidence for the details we have discussed already. Let's continue our EDA further:

```
# Viewing the first 5 records in the dataset
head(getRatingMatrix(Jester5k),5)
```

This will result in the following output:

```
2.5 x 100 sparse Matrix of class "dgCMatrix"
   [[ suppressing 100 column names 'j1', 'j2', 'j3' ... ]]
u2841    7.91   9.17   5.34   8.16  -8.74   7.14   8.88  -8.25   5.87   6.21   7.72
6.12  -0.73   7.77  -5.83  -8.88   8.98
u15547  -3.20  -3.50  -9.56  -8.74  -6.36  -3.30   0.78   2.18  -8.40  -8.79  -7.04
-6.02   3.35  -4.61   3.64  -6.41  -4.13
u15221  -1.70   1.21   1.55   2.77   5.58   3.06   2.72  -4.66   4.51  -3.06   2.33
3.93   0.05   2.38  -3.64  -7.72   0.97
u15573  -7.38  -8.93  -3.88  -7.23  -4.90   4.13   2.57   3.83   4.37   3.16  -4.90
-5.78  -5.83   2.52  -5.24   4.51   4.37
u21505   0.10   4.17   4.90   1.55   5.53   1.50  -3.79   1.94   3.59   4.81  -0.68
-0.97  -6.46  -0.34  -2.14  -2.04  -2.57
u2841   -9.32  -9.08  -9.13  7.77   8.59   5.29   8.25   6.02   5.24   7.82   7.96
-8.88   8.25   3.64  -0.73   8.25   5.34  -7.77
u15547  -0.15  -1.84  -1.84  1.84  -1.21  -8.59  -5.19  -2.18   0.19   2.57  -5.78
1.07  -8.79   3.01   2.67  -9.22  -9.32   3.69
u15221   2.04   1.94   4.42  1.17   0.10  -5.10  -3.25   3.35   3.30  -1.70   3.16
-0.29   1.36   3.54   6.17  -2.72   3.11   4.81
u15573   4.95   5.49  -0.49  3.40  -2.14   5.29  -3.11  -4.56  -5.44  -6.89  -0.24
-5.15  -3.59  -8.20   2.18   0.39  -1.21  -2.62
u21505  -0.15   2.43   3.16  1.50   4.37  -0.10  -2.14   3.98   2.38   6.84  -0.68
0.87   3.30   6.21   5.78  -6.21  -0.78  -1.36
## number of ratings
print(nratings(Jester5k))
```

This will result in the following output:

```
362106## number of ratings per user
```

We will print the summary of the dataset using the following command:

```
print(summary(rowCounts(Jester5k)))
```

This will result in the following output:

```
   Min. 1st Qu.  Median    Mean 3rd Qu.    Max.
  36.00   53.00   72.00   72.42  100.00  100.00
```

We will now plot the histogram:

```
## rating distribution
hist(getRatings(Jester5k), main="Distribution of ratings")
```

This will result in the following output:

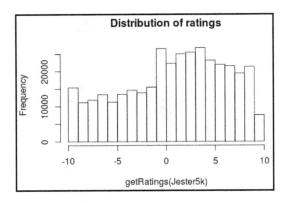

From the output, we see a somewhat normal distribution. It can also be seen that the positive ratings outnumber the negative ratings.

The `Jester5K` dataset also provides a character vector called `JesterJokes`. The vector is of length 100. These are the actual 100 jokes among others that were shown to the users to get the ratings. We could examine the jokes with the following command:

```
head(JesterJokes,5)
```

This will result in the following output:

```
j1 "A man visits the doctor. The doctor says \"I have bad news for you.You
have cancer and Alzheimer's disease\". The man replies \"Well,thank God I
don't have cancer!\""
j2 "This couple had an excellent relationship going until one day he came
home from work to find his girlfriend packing. He asked her why she was
leaving him and she told him that she had heard awful things about him.
\"What could they possibly have said to make you move out?\" \"They told me
that you were a pedophile.\" He replied, \"That's an awfully big word for a
ten year old.\""
j3  "Q. What's 200 feet long and has 4 teeth? A. The front row at a Willie
Nelson Concert."
j4 "Q. What's the difference between a man and a toilet? A. A toilet
doesn't follow you around after you use it."
j5 "Q. What's O. J. Simpson's Internet address? A. Slash, slash, backslash,
slash, slash, escape."
```

Based on the 5,000 user ratings we have, we could perform additional EDA to identify the joke that is rated as best by the users. This can be done through the following code:

```
## 'best' joke with highest average rating
best <- which.max(colMeans(Jester5k))
cat(JesterJokes[best])
```

This will result in the following output:

```
A guy goes into confession and says to the priest, "Father, I'm 80 years
old, widower, with 11 grandchildren. Last night I met two beautiful flight
attendants. They took me home and I made love to both of them. Twice." The
priest said: "Well, my son, when was the last time you were in confession?"
"Never Father, I'm Jewish." "So then, why are you telling me?" "I'm telling
everybody."
```

We could perform additional EDA to visualize the univariate and multivariate analysis. This exploration will help us understand each of the variables in detail as well as the relationship between them. While we do not delve deep into each of these aspects, here are some thoughts that can be explored:

- Exploring the users who always provide high ratings to most jokes
- Correlation between the ratings provided to jokes
- Identification of users that are very critical
- Exploring the most popular jokes or least popular jokes
- Identifying the jokes with the fewest ratings and identifying the associations between them

Converting the DataFrame

We are going to use functions from an R library called `recommenderlab` to build recommendation engine projects in this chapter. Irrespective of the category of recommendation system we implement, there are some prerequisites that the dataset needs to satisfy to be able to apply the `recommenderlab` functions. The prebuilt `recommenderlab` functions for collaborative filtering expects `realRatingMatrix` to be supplied as input. In our case, the `Jester5k` dataset is already in this format, therefore, we could directly use this matrix to apply the `recommenderlab` functions.

In case, we were to have our data as a R DataFrame and if we intend to convert into `realRatingMatrix`, the following steps may be performed:

1. Convert the DataFrame into an R matrix as follows:

```
# convert the df dataframe to a matrix
r_mat <- as.matrix(df)
```

2. Convert the resultant matrix into `realRatingMatrix` with the help of the `as()` function as follows:

```
# convert r_mat matrix to a recommenderlab realRatingMatrix
r_real_mat <- as(r_mat,"realRatingMatrix")
```

 Here, we assume that the name of the DataFrame is `df`, the code will convert it into a `realRatingMatrix` that can be used as input to the `recommenderlab` functions.

Dividing the DataFrame

Another prerequisite is to divide the dataset into train and test subsets. These subsets will be used in later sections to implement our recommendation systems and to measure the performance. The `evaluationScheme()` function from the `recommenderlab` library can be used to split the dataset into training and testing subsets. A number of user-specified parameters can be passed to this function. In the following code, `realRatingMatrix` is split according to an 80/20 training/testing split, with up to 20 items recommended for each user. Furthermore, we specify that any rating greater than 0 is to be considered a positive rating, in conformance with the predefined [-10, 10] rating scale. The `Jester5k` dataset can be divided into the train and test datasets with the following code:

```
# split the data into the training and the test set
Jester5k_es <- evaluationScheme(Jester5k, method="split", train=0.8,
given=20, goodRating=0)
# verifying if the train - test was done successfully
print(Jester5k_es)
```

This will result in the following output:

```
Evaluation scheme with 20 items given
Method: 'split' with 1 run(s).
Training set proportion: 0.800
Good ratings: >=0.000000
Data set: 5000 x 100 rating matrix of class 'realRatingMatrix' with 362106
ratings.
```

From the output of the `evaluationScheme()` function, we can observe that the function yielded a single R object containing both the training and test subsets. This object will be used to define and evaluate a variety of recommender models.

Building a recommendation system with an item-based collaborative filtering technique

The `recommenderlab` package of R offers the **item-based collaborative filtering (ITCF)** option to build a recommendation system. This is a very straightforward approach that just needs us to call the function and supply it with the necessary parameters. The parameters, in general, will have a lot of influence on the performance of the model; therefore, testing each parameter combination is the key to obtaining the best model for recommendations. The following are the parameters that can be passed to the `Recommender` function:

- **Data normalization**: Normalizing the ratings matrix is a key step in preparing the data for the recommendation engine. The process of normalization processes the ratings in the matrix by removing the rating bias. The possible values for this parameter are `NULL`, `Center`, and `Z-Score`.
- **Distance**: This represents the type of similarity metric to be used within the model. The possible values for this parameter are Cosine similarity, Euclidean distance, and Pearson's correlation.

With these parameter combinations, we could build and test 3 x 3 ITCF models. The basic intuition behind ITCF is that if a person likes item A, there is a good probability that they like item B as well, as long as items A and B are similar. It may be understood that the term *similar* does not indicate similarity between the items based on the item's attributes, but, a similarity in user preferences, for example, a group of people that liked items A also liked item B. The following diagram shows the working principle of ITCF:

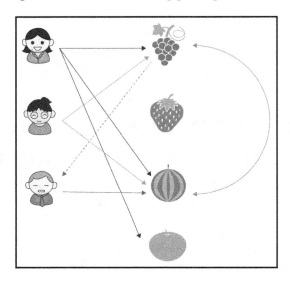

Example showing the working of item based collaborative filtering

Let's explore the diagram in a little more detail. In ITCF, the watermelon and grapes will form the similar-items neighborhood, which means that irrespective of users, different items that are equivalent will form a neighborhood. So when user X likes watermelon, the other item from the same neighborhood, which is grapes, will be recommended by the recommender system based on item-based collaborative filter.

ITCF involves the following three steps:

1. **Computing the item-based similarities through a distance measure**: This involves computing the distance between the items. The distance may be computed with one of the many distance measures, such as Cosine similarity, Euclidean distance, Manhattan distance, or Jaccard index. The output of this step is to obtain a similarity matrix where each cell corresponds to the similarity of the item specified on the row of the cell and the item specified on the column of the cell.

2. **Predicting the targeted item rating for a specific user**: The rating is arrived at by computing the weighted sum of ratings made to the item very similar to the target item.

3. **Recommending the top N items**: Once all the items are predicted, we recommend the top *N* items.

Now, let's build each one of the ITCF models and measure the performance against the test dataset. The following code trains the ITCF models with several parameter combinations:

```
type = "IBCF"
##train ITCF cosine similarity models
# non-normalized
ITCF_N_C <- Recommender(getData(Jester5k_es, "train"), type,
                        param=list(normalize = NULL, method="Cosine"))
# centered
ITCF_C_C <- Recommender(getData(Jester5k_es, "train"), type,
                        param=list(normalize = "center",method="Cosine"))
# Z-score normalization
ITCF_Z_C <- Recommender(getData(Jester5k_es, "train"), type,
                        param=list(normalize = "Z-score",method="Cosine"))
##train ITCF Euclidean Distance models
# non-normalized
ITCF_N_E <- Recommender(getData(Jester5k_es, "train"), type,
                        param=list(normalize = NULL, method="Euclidean"))
# centered
ITCF_C_E <- Recommender(getData(Jester5k_es, "train"), type,
                        param=list(normalize =
"center",method="Euclidean"))
# Z-score normalization
ITCF_Z_E <- Recommender(getData(Jester5k_es, "train"), type,
                        param=list(normalize = "Z-
score",method="Euclidean"))
#train ITCF pearson correlation models
# non-normalized
ITCF_N_P <- Recommender(getData(Jester5k_es, "train"), type,
                        param=list(normalize = NULL, method="pearson"))
# centered
ITCF_C_P <- Recommender(getData(Jester5k_es, "train"), type,
                        param=list(normalize = "center",method="pearson"))
# Z-score normalization
ITCF_Z_P <- Recommender(getData(Jester5k_es, "train"), type,
                        param=list(normalize = "Z-score",method="pearson"))
```

We now have the ITCF models, so let's get to computing the performance on the test data with each of the models we have created. The objective is to identify the best-performing ITCF model for this dataset. The following code gets the performance measurements with all the nine models on the test dataset:

```
# compute predicted ratings from each of the 9 models on the test dataset
pred1 <- predict(ITCF_N_C, getData(Jester5k_es, "known"), type="ratings")
pred2 <- predict(ITCF_C_C, getData(Jester5k_es, "known"), type="ratings")
pred3 <- predict(ITCF_Z_C, getData(Jester5k_es, "known"), type="ratings")
pred4 <- predict(ITCF_N_E, getData(Jester5k_es, "known"), type="ratings")
pred5 <- predict(ITCF_C_E, getData(Jester5k_es, "known"), type="ratings")
pred6 <- predict(ITCF_Z_E, getData(Jester5k_es, "known"), type="ratings")
pred7 <- predict(ITCF_N_P, getData(Jester5k_es, "known"), type="ratings")
pred8 <- predict(ITCF_C_P, getData(Jester5k_es, "known"), type="ratings")
pred9 <- predict(ITCF_Z_P, getData(Jester5k_es, "known"), type="ratings")
# set all predictions that fall outside the valid range to the boundary
values
pred1@data@x[pred1@data@x[] < -10] <- -10
pred1@data@x[pred1@data@x[] > 10] <- 10
pred2@data@x[pred2@data@x[] < -10] <- -10
pred2@data@x[pred2@data@x[] > 10] <- 10
pred3@data@x[pred3@data@x[] < -10] <- -10
pred3@data@x[pred3@data@x[] > 10] <- 10
pred4@data@x[pred4@data@x[] < -10] <- -10
pred4@data@x[pred4@data@x[] > 10] <- 10
pred5@data@x[pred5@data@x[] < -10] <- -10
pred5@data@x[pred5@data@x[] > 10] <- 10
pred6@data@x[pred6@data@x[] < -10] <- -10
pred6@data@x[pred6@data@x[] > 10] <- 10
pred7@data@x[pred7@data@x[] < -10] <- -10
pred7@data@x[pred7@data@x[] > 10] <- 10
pred8@data@x[pred8@data@x[] < -10] <- -10
pred8@data@x[pred8@data@x[] > 10] <- 10
pred9@data@x[pred9@data@x[] < -10] <- -10
pred9@data@x[pred9@data@x[] > 10] <- 10
# aggregate the performance measurements obtained from all the models
error_ITCF <- rbind(
  ITCF_N_C = calcPredictionAccuracy(pred1, getData(Jester5k_es,
"unknown")),
  ITCF_C_C = calcPredictionAccuracy(pred2, getData(Jester5k_es,
"unknown")),
  ITCF_Z_C = calcPredictionAccuracy(pred3, getData(Jester5k_es,
"unknown")),
  ITCF_N_E = calcPredictionAccuracy(pred4, getData(Jester5k_es,
"unknown")),
  ITCF_C_E = calcPredictionAccuracy(pred5, getData(Jester5k_es,
"unknown")),
  ITCF_Z_E = calcPredictionAccuracy(pred6, getData(Jester5k_es,
```

```
"unknown")),
  ITCF_N_P = calcPredictionAccuracy(pred7, getData(Jester5k_es,
"unknown")),
  ITCF_C_P = calcPredictionAccuracy(pred8, getData(Jester5k_es,
"unknown")),
  ITCF_Z_P = calcPredictionAccuracy(pred9, getData(Jester5k_es, "unknown"))
)
library(knitr)
kable(error_ITCF)
```

This will result in the following output:

```
|          |     RMSE|      MSE|      MAE|
|:---------|--------:|--------:|--------:|
|ITCF_N_C  | 4.533455| 20.55221| 3.460860|
|ITCF_C_C  | 5.082643| 25.83326| 4.012391|
|ITCF_Z_C  | 5.089552| 25.90354| 4.021435|
|ITCF_N_E  | 4.520893| 20.43848| 3.462490|
|ITCF_C_E  | 4.519783| 20.42844| 3.462271|
|ITCF_Z_E  | 4.527953| 20.50236| 3.472080|
|ITCF_N_P  | 4.582121| 20.99583| 3.522113|
|ITCF_C_P  | 4.545966| 20.66581| 3.510830|
|ITCF_Z_P  | 4.569294| 20.87845| 3.536400|
```

We see the output that the ITCF recommendation application on data with the Euclidean distance yielded the best performance measurement.

Building a recommendation system with a user-based collaborative filtering technique

The Jokes recommendation system we built earlier, with item-based filtering, uses the powerful `recommenderlab` library available in R. In this implementation of the **user-based collaborative filtering (UBCF)** approach, we make use of the same library.

The following diagram shows the working principle of UBCF:

Example depicting working principle of user based collaborative filter

To understand the concept better, let's discuss the preceding diagram in detail. Let's assume that there are three users: X,Y, and Z. In UBCF, users X and Z are very similar as both of them like strawberries and watermelons. User X also likes grapes and oranges. So a user-based collaborative filter recommends grapes and oranges to user Z. The idea is that similar people tend to like similar things.

The primary difference between a user-based collaborative filter and an item-based collaborative filter is demonstrated by the following recommendation captions often seen in online retail sites:

- **ITCF**: Customers who bought this item also bought
- **UBCF**: Customers similar to you bought

A user-based collaborative filter is built upon the following three key steps:

1. Find the **k-nearest neighbors** (**KNN**) to the user *x*, using a similarity function, *w*, to measure the distance between each pair of users:

$$similarity(x, i) = w(x, i), \ i \ is \ a \ member \ of \ k$$

2. Predict the rating that user *x* will provide to all items the KNN has rated, but *x* has not.

3. The *N* recommended items to user *x* is the top *N* items that have the best predicted ratings.

In short, a user-item matrix is constructed during the UBCF process and based on similar users, the ratings of the unseen items of a user are predicted. The items that get the highest ratings among the predictions form the final list of recommendations.

The implementation of this project is very similar to ITCF as we are using the same library. The only change required in the code is to change the IBCF method to use UBCF. The following code block is the full code of the project implementation with UBCF:

```
library(recommenderlab)
data(Jester5k)
# split the data into the training and the test set
Jester5k_es <- evaluationScheme(Jester5k, method="split", train=0.8,
given=20, goodRating=0)
print(Jester5k_es)
type = "UBCF"
#train UBCF cosine similarity models
# non-normalized
UBCF_N_C <- Recommender(getData(Jester5k_es, "train"), type,
                        param=list(normalize = NULL, method="Cosine"))
# centered
UBCF_C_C <- Recommender(getData(Jester5k_es, "train"), type,
                        param=list(normalize = "center",method="Cosine"))
# Z-score normalization
UBCF_Z_C <- Recommender(getData(Jester5k_es, "train"), type,
                        param=list(normalize = "Z-score",method="Cosine"))
#train UBCF Euclidean Distance models
# non-normalized
UBCF_N_E <- Recommender(getData(Jester5k_es, "train"), type,
                        param=list(normalize = NULL, method="Euclidean"))
# centered
UBCF_C_E <- Recommender(getData(Jester5k_es, "train"), type,
                        param=list(normalize =
"center",method="Euclidean"))
# Z-score normalization
UBCF_Z_E <- Recommender(getData(Jester5k_es, "train"), type,
                        param=list(normalize = "Z-
score",method="Euclidean"))
#train UBCF pearson correlation models
# non-normalized
UBCF_N_P <- Recommender(getData(Jester5k_es, "train"), type,
                        param=list(normalize = NULL, method="pearson"))
# centered
```

```
UBCF_C_P <- Recommender(getData(Jester5k_es, "train"), type,
                        param=list(normalize = "center",method="pearson"))
# Z-score normalization
UBCF_Z_P <- Recommender(getData(Jester5k_es, "train"), type,
                        param=list(normalize = "Z-score",method="pearson"))
# compute predicted ratings from each of the 9 models on the test dataset
pred1 <- predict(UBCF_N_C, getData(Jester5k_es, "known"), type="ratings")
pred2 <- predict(UBCF_C_C, getData(Jester5k_es, "known"), type="ratings")
pred3 <- predict(UBCF_Z_C, getData(Jester5k_es, "known"), type="ratings")
pred4 <- predict(UBCF_N_E, getData(Jester5k_es, "known"), type="ratings")
pred5 <- predict(UBCF_C_E, getData(Jester5k_es, "known"), type="ratings")
pred6 <- predict(UBCF_Z_E, getData(Jester5k_es, "known"), type="ratings")
pred7 <- predict(UBCF_N_P, getData(Jester5k_es, "known"), type="ratings")
pred8 <- predict(UBCF_C_P, getData(Jester5k_es, "known"), type="ratings")
pred9 <- predict(UBCF_Z_P, getData(Jester5k_es, "known"), type="ratings")
# set all predictions that fall outside the valid range to the boundary
values
pred1@data@x[pred1@data@x[] < -10] <- -10
pred1@data@x[pred1@data@x[] > 10] <- 10
pred2@data@x[pred2@data@x[] < -10] <- -10
pred2@data@x[pred2@data@x[] > 10] <- 10
pred3@data@x[pred3@data@x[] < -10] <- -10
pred3@data@x[pred3@data@x[] > 10] <- 10
pred4@data@x[pred4@data@x[] < -10] <- -10
pred4@data@x[pred4@data@x[] > 10] <- 10
pred5@data@x[pred5@data@x[] < -10] <- -10
pred5@data@x[pred5@data@x[] > 10] <- 10
pred6@data@x[pred6@data@x[] < -10] <- -10
pred6@data@x[pred6@data@x[] > 10] <- 10
pred7@data@x[pred7@data@x[] < -10] <- -10
pred7@data@x[pred7@data@x[] > 10] <- 10
pred8@data@x[pred8@data@x[] < -10] <- -10
pred8@data@x[pred8@data@x[] > 10] <- 10
pred9@data@x[pred9@data@x[] < -10] <- -10
pred9@data@x[pred9@data@x[] > 10] <- 10
# aggregate the performance statistics
error_UBCF <- rbind(
  UBCF_N_C = calcPredictionAccuracy(pred1, getData(Jester5k_es,
"unknown")),
  UBCF_C_C = calcPredictionAccuracy(pred2, getData(Jester5k_es,
"unknown")),
  UBCF_Z_C = calcPredictionAccuracy(pred3, getData(Jester5k_es,
"unknown")),
  UBCF_N_E = calcPredictionAccuracy(pred4, getData(Jester5k_es,
"unknown")),
  UBCF_C_E = calcPredictionAccuracy(pred5, getData(Jester5k_es,
"unknown")),
  UBCF_Z_E = calcPredictionAccuracy(pred6, getData(Jester5k_es,
```

```
"unknown")),
   UBCF_N_P = calcPredictionAccuracy(pred7, getData(Jester5k_es,
"unknown")),
   UBCF_C_P = calcPredictionAccuracy(pred8, getData(Jester5k_es,
"unknown")),
   UBCF_Z_P = calcPredictionAccuracy(pred9, getData(Jester5k_es, "unknown"))
)
library(knitr)
print(kable(error_UBCF))
```

This will result in the following output:

	RMSE	MSE	MAE
UBCF_N_C	4.877935	23.79425	3.986170
UBCF_C_C	4.518210	20.41422	3.578551
UBCF_Z_C	4.517669	20.40933	3.552120
UBCF_N_E	4.644877	21.57488	3.778046
UBCF_C_E	4.489157	20.15253	3.552543
UBCF_Z_E	4.496185	20.21568	3.528534
UBCF_N_P	4.927442	24.27968	4.074879
UBCF_C_P	4.487073	20.13382	3.553429
UBCF_Z_P	4.484986	20.11510	3.525356

Based on the UBCF output, we observe that the $z-score$ normalized data with Pearson's correlation as the distance has yielded the best performance measurement. Furthermore, if we want, the UBCF and ITCF results may be compared (testing needs to be done on the same test dataset) to arrive at a conclusion of accepting the best model among the 18 models that are built for the final recommendation engine deployment.

The key point to observe in the code is the UBCF value that is passed to the method parameter. In the previous project, we built an item-based collaborative filter; all that is needed is for us to replace the value passed to the method parameter with IBCF.

Building a recommendation system based on an association-rule mining technique

Association-rule mining, or market-basket analysis, is a very popular data mining technique used in the retail industry to identify the products that need to be kept together so as to encourage cross sales. An interesting aspect behind this algorithm is that historical invoices are mined to identify the products that are bought together.

There are several off-the-shelf algorithms available to perform market-basket analysis. Some of them are Apriori, **equivalence class transformation** (**ECLAT**), and **frequent pattern growth** (**FP-growth**). We will learn to solve our problem of recommending jokes to users through applying the Apriori algorithm on the Jester jokes dataset. We will now learn the theoretical aspects that underpin the Apriori algorithm.

The Apriori algorithm

The building blocks of the algorithm are the items that are found in any given transaction. Each transaction could have one or more items in it. The items that form a transaction are called an itemset. An example of a transaction is an invoice.

Given the transactions dataset, the objective is to find the items in data that are associated with each other. Association is measured as frequency of the occurrence of the items in the same context. For example, purchasing one product when another product is purchased represents an association rule. The association rule detects the common usage of items.

More formally, we can define association-rule mining as, given a set of items I = {I1,I2,..Im} and database of transactions D = {t1,t,2..tn}, where ti= { Ii1,Ii2..Iim} where Iik is element of, an association is an implication of X->Y where X,Y subset of I are set of items and X intersection Y is φ. In short, associations express an implication from X-> Y, where X and Y are itemsets.

The algorithm can be better understood by an example. So, let's consider the following table, which shows a representative list of sample transactions in a supermarket:

Transaction	Items
1	Milk, curd, chocolate
2	Bread, butter
3	Coke, jam
4	Bread, milk, butter, Coke
5	Bread, milk, butter, jam

Sample transactions in a super market

Let's try to explore some fundamental concepts that will help us understand how the Apriori algorithm works:

- **Item**: An item is any individual product that is part of each of the transactions. For example, milk, Coke, and butter are all termed as items.
- **Itemset**: Collection of one or more items. For example, *{butter, milk, coke}*, *{butter, milk}*.

- **Support count**: Frequency of occurrence of an itemset. For example, support count or σ {butter, bread, milk} = 2.
- **Support**: A fraction of transactions that contain an itemset. For example, s = {butter, bread, milk} = 2/5.
- **Frequent itemset**: An itemset whose support is greater than the minimum threshold.
- **Support for an itemset in a context**: Fraction of contexts that contain both X and Y:

$$s = Support\_count(X \bigcup Y)/N$$

So, s for {milk, butter} -> {bread} will be $s = \sigma$ {milk, butter, bread}/N = 2/5 = 0.4

- **Confidence**: Measures the strength of the rule, whereas support measures how often it should occur in the database. It computes how often items in Y occur in containing X through the following formula:

$$c = Support\_count(X \bigcup Y)/Support\_count(X)$$

For example: For {bread} -> {butter}

c or $\alpha = \sigma$ {butter, bread} / σ {bread} = 3/3 = 1

Let's consider another example confidence for {curd} -> {bread}:

c or $\alpha = \sigma$ {curd,bread} / σ {bread} = 0/3 = 0

The Apriori algorithm intends to generate all possible combinations of the itemsets from the list of the items and then prunes the itemsets that have met the predefined support and confidence parameter values that were passed to the algorithm. So, it may be understood that the Apriori algorithm is a two-step algorithm:

1. Generating itemsets from the items
2. Evaluating and pruning the itemsets based on predefined support and confidence

Let's discuss step 1 in detail. Assume there are n items in the collection. The number of itemsets one could create is 2^n, and all these need to be evaluated in the second step in order to come up with the final results. Even considering just 100 different items, the number of itemsets generated is 1.27e+30! The huge number of itemsets poses a severe computational challenge.

The Apriori algorithm overcomes this challenge by preempting the itemsets that are generally rare or less important. The Apriori principle states that *if an itemset is frequent, all of its subsets must also be frequent.* This means that if an item does not meet the predefined support threshold, then such item does not participate in the creation of itemsets. The Apriori algorithm thus comes up with restricted number of itemsets that are viable to be evaluated without encountering a computational challenge.

The first step of the algorithm is iterative in nature. In the first iteration, it considers all itemsets of length 1, that is, each itemset contains only one item in it. Then each item is evaluated to eliminate the itemsets that are found to not meet the preset support threshold. The output of the first iteration is all itemsets of length 1 that meet the required support. This becomes the input for iteration 2, and now itemsets of length 2 are formed using only the final itemsets that are output in first iteration. Each of the itemsets formed during step 2 is checked again for the support threshold; if it is not met, such itemsets are eliminated. The iterations continue until no new itemsets can be created. The process of itemsets is illustrated in the following diagram:

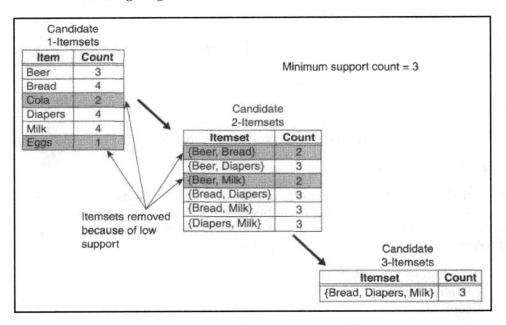

Illustration showing the itemsets creation in Apriori algorithm

Once we have all itemsets post all the step 1 iterations of the algorithm, step 2 kicks in. Each of the itemsets generated is tested to check whether it meets the predefined confidence value. If it does not meet the threshold, such itemsets are eliminated from the final output.

At a stage where all iterations are complete and the final rules are the output from Apriori, we make use of a metric called lift to consume the relevant rules from the final output. Lift defines how much more likely one item or itemset is purchased relative to its typical rate of purchase, given that we know another item or itemset has been purchased. For each itemset, we get the lift measurement using the following formula:

$$Lift(X->Y) = Confidence(X->Y)/Support(Y)$$

Let's delve a little deeper into understanding the lift metric. Assume that in a supermarket, milk and bread are bought together by chance. In such a case, a large number of transactions are expected to cover the milk and bread purchased. A lift (milk -> bread) of more than 1 implies that these items are found together more often than these items are purchased together by chance. We generally would look for lift values greater than 1 when evaluating the rules for their usefulness in business. A lift value higher than 1 indicates that the itemset generated is very strong, and therefore worth considering for implementation.

Now, let's implement the recommendation system using the Apriori algorithm:

```
# load the required libraries
library(data.table)
library(arules)
library(recommenderlab)
# set the seed so that the results are replicable
set.seed(42)
# reading the Jester5k data
data(Jester5k)
class(Jester5k)
```

This will result in the following output:

```
[1] "realRatingMatrix"
attr(,"package")
[1] "recommenderlab"
```

We can see from the output that the `Jester5k` data in the `recommenderlab` library is in the `realRatingsMatrix` format. We are also aware that the cells in this matrix contain the ratings provided by the users for various jokes and we are aware that the ratings range between -10 to +10.

Applying the Apriori algorithm on the `Jester5k` dataset give us an opportunity to understand the association between the jokes. However, prior to applying the Apriori algorithm, we will need to transform the dataset to binary values where 1 represents a positive rating and 0 represents a negative rating or no rating. The `recommenderlab` library comes up with the `binarize()` function, which can perform the required operation for us. The following code binarizes the ratings matrix:

```
# binarizing the Jester ratings
Jester5k_bin <- binarize(Jester5k, minRating=1)
# let us verify the binarized object
class(Jester5k_bin)
```

This will result in the following output:

```
[1] "binaryRatingMatrix"
attr(,"package")
[1] "recommenderlab"
```

We can observe from the output that `realRatingsMatrix` is successfully converted into `binaryRatingMatrix`. The Apriori algorithm that mines the associations expects a matrix to be passed as input rather than `binaryRatingMatrix`. We can very easily convert the `Jester5k_bin` object to the matrix format with the following code:

```
# converting the binaryratingsmatrix to matrix format
Jester5k_bin_mat <- as(Jester5k_bin, "matrix")
# visualizing the matrix object
View(Jester5k_bin_mat)
```

This will result in the following output:

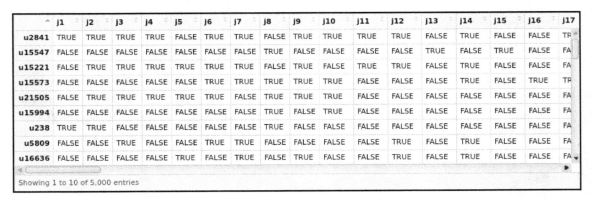

	j1	j2	j3	j4	j5	j6	j7	j8	j9	j10	j11	j12	j13	j14	j15	j16	j17
u2841	TRUE	TRUE	TRUE	TRUE	FALSE	TRUE	TRUE	FALSE	TRUE	TRUE	TRUE	TRUE	FALSE	TRUE	FALSE	FALSE	TR
u15547	FALSE	FALSE	FALSE	FALSE	FALSE	FALSE	FALSE	TRUE	FALSE	FALSE	FALSE	FALSE	TRUE	FALSE	TRUE	FALSE	FA
u15221	FALSE	TRUE	TRUE	TRUE	TRUE	TRUE	TRUE	FALSE	TRUE	FALSE	TRUE	TRUE	FALSE	TRUE	FALSE	FALSE	FA
u15573	FALSE	FALSE	FALSE	FALSE	FALSE	TRUE	TRUE	TRUE	TRUE	TRUE	FALSE	FALSE	FALSE	TRUE	FALSE	TRUE	TR
u21505	FALSE	TRUE	TRUE	TRUE	TRUE	TRUE	FALSE	TRUE	TRUE	TRUE	FALSE	FALSE	FALSE	FALSE	FALSE	FALSE	FA
u15994	FALSE	FALSE	FALSE	FALSE	FALSE	FALSE	FALSE	TRUE	FALSE	TRUE	FALSE	FALSE	FALSE	FALSE	FALSE	FALSE	FA
u238	TRUE	TRUE	FALSE	FALSE	FALSE	FALSE	FALSE	TRUE	FALSE	FALSE	FALSE	FALSE	FALSE	FALSE	FALSE	FALSE	FA
u5809	FALSE	FALSE	TRUE	FALSE	FALSE	TRUE	TRUE	FALSE	FALSE	FALSE	FALSE	TRUE	FALSE	TRUE	FALSE	FALSE	FA
u16636	FALSE	FALSE	FALSE	FALSE	TRUE	FALSE	TRUE	FALSE	TRUE	FALSE	FALSE	TRUE	FALSE	TRUE	FALSE	FALSE	FA

Showing 1 to 10 of 5,000 entries

We see from the output that all the cells of the matrix are represented as TRUE and FALSE, but Apriori expects the cells to be numeric. Let's now convert the cells into 1 and 0 for TRUE and FALSE, respectively, with the following code:

```
# converting the cell values to 1 and 0
Jester5k_bin_mat_num <- 1*Jester5k_bin_mat
# viewing the matrix
View(Jester5k_bin_mat_num)
```

This will result in the following output:

	j1	j2	j3	j4	j5	j6	j7	j8	j9	j10	j11	j12	j13	j14	j15	j16	j17
u2841	1	1	1	1	0	1	1	0	1	1	1	1	0	1	0	0	
u15547	0	0	0	0	0	0	0	1	0	0	0	0	1	0	1	0	
u15221	0	1	1	1	1	1	1	0	1	0	1	1	0	1	0	0	
u15573	0	0	0	0	0	1	1	1	1	1	0	0	0	1	0	1	
u21505	0	1	1	1	1	1	0	1	1	1	0	0	0	0	0	0	
u15994	0	0	0	0	0	0	0	1	0	1	0	0	0	0	0	0	
u238	1	1	0	0	0	0	0	1	0	0	0	0	0	0	0	0	
u5809	0	0	1	0	0	1	1	0	0	0	0	1	0	1	0	0	
u16636	0	0	0	0	1	0	1	0	1	0	0	1	0	1	0	0	

Showing 1 to 10 of 5,000 entries

Now we are all set to apply the Apriori algorithm on the dataset. There are two parameters, support and confidence, that we need to pass to the algorithm. The algorithm mines the dataset based on these two parameter values. We pass 0.5 as the value for support and 0.8 as the value for confidence. The following line of code extracts the joke associations that exist in our Jester jokes dataset:

```
rules <- apriori(data = Jester5k_bin_mat_num, parameter = list(supp =
0.005, conf = 0.8))
```

This will result in the following output:

```
Apriori
Parameter specification:
 confidence minval smax arem  aval originalSupport maxtime support minlen
maxlen target    ext
       0.8    0.1    1 none FALSE            TRUE       5     0.5      1
10   rules FALSE
Algorithmic control:
 filter tree heap memopt load sort verbose
    0.1 TRUE TRUE   FALSE TRUE    2     TRUE
Absolute minimum support count: 2500
set item appearances ...[0 item(s)] done [0.00s].
```

```
set transactions ...[100 item(s), 5000 transaction(s)] done [0.02s].
sorting and recoding items ... [29 item(s)] done [0.00s].
creating transaction tree ... done [0.00s].
checking subsets of size 1 2 3 done [0.01s].
writing ... [78 rule(s)] done [0.00s].
creating S4 object  ... done [0.00s].
```

The `rules` object that was created from the execution of the Apriori algorithm now has all the joke associations that were extracted and mined from the dataset. As we can see from the output, there are 78 jokes associations that were extracted in total. We can examine the rules with the following line of code:

```
inspect(rules)
```

This will result in the following output:

```
    lhs        rhs    support confidence lift      count
[1] {j48}   => {j50} 0.5068  0.8376860  1.084523 2534
[2] {j56}   => {j36} 0.5036  0.8310231  1.105672 2518
[3] {j56}   => {j50} 0.5246  0.8656766  1.120762 2623
[4] {j42}   => {j50} 0.5150  0.8475971  1.097355 2575
[5] {j31}   => {j27} 0.5196  0.8255481  1.146276 2598
```

The output shown is just five rules out of the overall 78 rules that are in the list. The way to read each rule is that the joke shown on the left column (`lhs`) leads to the joke on the right column (`rhs`); that is, a user that liked the joke on `lhs` of the rule generally tends to like the joke shown on `rhs`. For example, in the first rule, if a user has liked joke j48, it is likely that they will also like j50, therefore it is worth recommending joke j50 to the user that has only read joke j48.

While there are several rules generated by the Apriori algorithm, the strength of each rule is specified by a metric, called `lift`. This is a metric that describes the worthiness of a rule in a business context. Note that for a rule to be considered general, it has to have a lift that is less than or equal to 1. A lift value greater than 1 signifies a better rule for implementing in business. The aim of the following lines of code is to get such strong rules to the top of the list:

```
# converting the rules object into a dataframe
rulesdf <- as(rules, "data.frame")
# employing quick sort on the rules dataframe. lift and confidence are
# used as keys to sort the dataframe. - in the command indicates that we
# want lift and confidence to be sorted in descending order
rulesdf[order(-rulesdf$lift, -rulesdf$confidence), ]
```

This will result in the following output:

```
              rules support confidence     lift count
59 {j29,j50} => {j35}  0.5024  0.8348288 1.167266  2512
60 {j35,j50} => {j29}  0.5024  0.8301388 1.160709  2512
71 {j50,j53} => {j32}  0.5070  0.8385710 1.154420  2535
24      {j68} => {j62}  0.5478  0.8096364 1.149725  2739
72 {j32,j50} => {j53}  0.5070  0.8278903 1.149528  2535
5       {j31} => {j27}  0.5196  0.8255481 1.146276  2598
36      {j49} => {j62}  0.5578  0.8030521 1.140375  2789
78 {j36,j50} => {j32}  0.5220  0.8277831 1.139569  2610
68 {j27,j50} => {j36}  0.5128  0.8563794 1.139408  2564
66 {j36,j50} => {j35}  0.5132  0.8138281 1.137903  2566
58 {j29,j35} => {j50}  0.5024  0.8770950 1.135545  2512
32      {j69} => {j53}  0.5550  0.8171378 1.134598  2775
77 {j32,j50} => {j36}  0.5220  0.8523841 1.134093  2610
76 {j32,j36} => {j50}  0.5220  0.8755451 1.133538  2610
73 {j36,j53} => {j50}  0.5112  0.8747433 1.132500  2556
70 {j32,j53} => {j50}  0.5070  0.8747412 1.132498  2535
64 {j35,j36} => {j50}  0.5132  0.8745740 1.132281  2566
69 {j36,j50} => {j27}  0.5128  0.8131938 1.129122  2564
```

It may be observed that the output shown is only a subset of the rules output. The first rule indicates that j35 is a joke that can be recommended to a user that has already read jokes j29 and j50.

Likewise, we could just write a script to search all the jokes that a user has already read and match it with the left side of the rule; if a match is found, the corresponding right side of the rule can be recommended as the joke for the user.

Content-based recommendation engine

A recommendation engine that is solely based on the explicit or implicit feedback received from customers is termed as **content-based recommendation system**. Explicit feedback is the customer's expression of the interest through filling in a survey about preferences or rating jokes of interest or opting for newsletters related to the joke or adding the joke on the watchlist, and so on. Implicit feedback is more of a mellowed-out approach where a customer visits a page, clicks on a joke link, or just spends time reading a joke review on an e-commerce page. Based on the feedback received, similar jokes are recommended to the customers. It may be noted that content-based recommendations do not take into consideration the preferences and feedback of other customers in the system; instead, it is purely based on the personalized feedback from the specific customer.

In the recommendation process, the system identifies the products that are already positively rated by the customer with the products that the customer has not rated and looks for equivalents. Products that are similar to the positively-rated ones are recommended to the customers. In this model, the customer's preferences and behavior play a major role in incrementally fine-tuning the recommendations—that is, with each recommendation and based on whether the customer responded to the recommendation, the system learns incrementally to recommend differently. The following diagram is an illustration of how a content-based recommendation system works:

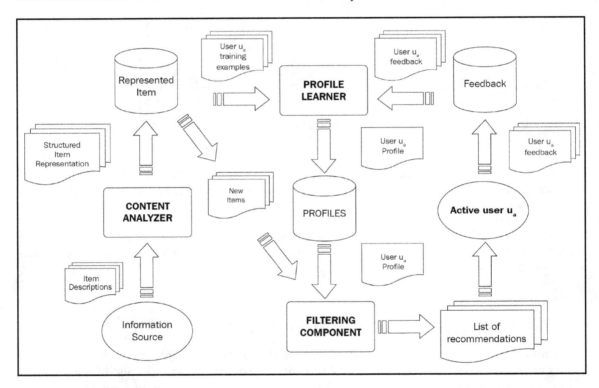

Working of a content based recommendation system

In our Jester jokes dataset, we have ratings given by users for various jokes as well as the content of the jokes themselves. Remember that the `JesterJokes` character vector incorporates the text present in the jokes themselves. Similarities between the texts present in the jokes can be used as one method to recommend jokes to users. The assumption is that if a person liked the content in a joke, and if there is another joke whose content is very similar, recommending the latter joke is probably going to be liked by the user.

Additional metadata related to jokes is not given in the Jester jokes dataset, however such metadata may be created from the content of the jokes. For example, the length of the joke, number of nouns, number of funny terms present in the joke, and central theme in the joke. Processing the text is not purely a recommendation area but it involves using NLP techniques as well. As we will be covering NLP in a different chapter, we will not cover it here.

Differentiating between ITCF and content-based recommendations

It might appear that item-based collaborative and content-based recommendations are the same. In reality, they are not the same. Let's touch upon the differences.

ITCF is totally based on user-item rankings. When we compute the similarity between items, we do not include the item attributes and just compute the similarity of items based on all customers' ratings. So the similarity between items is computed based on the ratings instead of the metadata of item itself.

In content-based recommendations, we make use of the content of both the user and the item. Generally, we construct a user profile and item profile using the content of a shared attribute space. For example, for a movie, we represent it with the actors in it and the genre (using binary coding, for example). For a user profile, we can do the same thing based on the user, such as some actors/genres. Then the similarity of user and item can be computed using cosine similarity, for example. This cosine measure leads to the recommendations.

Content-based filtering identifies products that are similar based on the tags assigned to each product. Each product is assigned weights on the basis of term frequency and inverse document frequency of each tag. After this, the user's probability of liking a product is calculated in order to arrive at the final recommendation list.

While content-based recommendation systems are highly efficient and personalized, there is an inherent problem with this model. Let's understand the over-specialization problem of content-based recommendations with an example.

Assume there are the following five movie genres:

- Comedy
- Thriller
- Science fiction
- Action
- Romance

There is this customer, Jake, who generally watches thriller and science fiction movies. Based on this preference, the content-based recommendation engine will only recommend movies related to these genres and it is never going to recommend movies from other categories. This problem arises as content-based recommendation engine solely relies on the user's past behavior and preferences to determine the recommendation.

Unlike content-recommendation systems, in ITCF recommendations, similar products build neighborhoods based on positive preferences of customers. Therefore, the system generates recommendations with products in the neighborhood that a customer might prefer. ITCF does this by making use of the correlation between the items based on the ratings given them by different users, while collaborative filtering relies on past preferences or rating correlation between users and it is able to generate recommendations for similar products even from customer's interest domain. This technique can lead to bad predictions if the product is unpopular and very few users have given feedback about it.

Building a hybrid recommendation system for Jokes recommendations

We see that both content-based filtering and collaborative filtering have their strengths and weaknesses. To overcome the issues, organizations build recommender systems that combine two or more technique and they are termed hybrid recommendation models. An example of this is a combination of content-based, IBCF, UBCF, and model-based recommender engine. This takes into account all the possible aspects that contribute to making the most relevant recommendation to the user. The following diagram shows an example approach followed in hybrid recommendation engines:

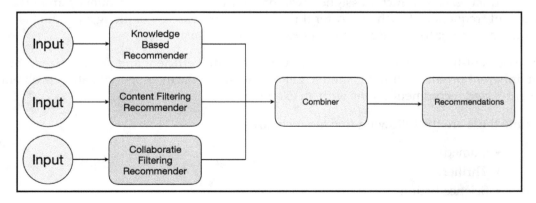

Sample approach to hybrid recommendation engine

We need to note that there is no standard approach to achieving a hybrid recommendation engine. In order to combine recommendations, here are some suggested strategies:

- **Voting**: Apply voting among the recommendation output obtained from individual recommender systems.
- **Rules-based selection**: We could devise rules that suggest weighting the output recommendations obtained from individual recommender systems. In this case, the output from recommender systems that got higher weights will be dominant and have more influence on the final recommendation outcome.
- **Combination**: Recommendations from all the recommender engines are presented together. A final list of recommendations is just the union of all recommendation output obtained from individual recommender systems.
- **Attribute integration**: Taking metadata from all recommender system to infuse it as input to another recommender.

Again, what works for a problem may not work for another, therefore these strategies need to be tested individually prior to coming up with final recommendation strategy.

The `recommenderlab` library offers the `HybridRecommender` function which allows users to train multiple recommender engines on the same set of data in one go and combine the predictions. The function has a weights parameter that offers a way to specify the weight of each of the models that will be used to combine individual predictions to arrive at the final recommendation predictions on unseen data. Implementing a hybrid recommendation-engine-based project is super straightforward and not too different from the code we learned in item-based collaborative filtering or user-based collaborative filtering projects. Anyway, let's write the code and build a hybrid recommendation engine for the `Jester5k` dataset:

```
# including the required libraries
library(recommenderlab)
# accessing the Jester5k dataset that is a part of recommenderlab library
data(Jester5k)
# split the data into the training and the test set
Jester5k_es <- evaluationScheme(Jester5k, method="split", train=0.8,
given=20, goodRating=0)
```

The preceding code is what trains a hybrid recommender. This is where it differs from the ITCF or UBCF recommenders we've built. We can observe from the code that we have used four different recommender methods that will constitute the hybrid recommender. Let's discuss each of these methods:

- The popular recommendation method simply recommends the popular jokes (determined by the number of ratings received) to users.
- The second recommender method we have used is item-based collaborative filtering method with non-normalized data but with distance being computed between items through cosine similarity.
- User-based collaborative filtering on Z-score normalized data with Euclidean distance being computed between users in the data.
- A random recommendation method that provides a random recommendation to the users.

By no means, we finalize that the combination of these four recommender methods is the best hybrid for this problem. The intention of this project is to demonstrate the implementation of the hybrid recommender. The choice of the methods involved is purely arbitrary. In reality, we may need to try multiple combinations to identify the best hybrid. The hybrid classifier is built using the following code:

```
#train a hybrid recommender model
hybrid_recom <- HybridRecommender(
  Recommender(getData(Jester5k_es, "train"), method = "POPULAR"),
  Recommender(getData(Jester5k_es, "train"), method="IBCF",
              param=list(normalize = NULL, method="Cosine")),
  Recommender(getData(Jester5k_es, "train"), method="UBCF",
                        param=list(normalize = "Z-
score",method="Euclidean")),
  Recommender(getData(Jester5k_es, "train"), method = "RANDOM"),
  weights = c(.2, .3, .3,.2)
)
# Observe the model that is built
print (getModel(hybrid_recom)
```

This will result in the following output:

```
$recommender
$recommender[[1]]
Recommender of type 'POPULAR' for 'realRatingMatrix'
learned using 4000 users.
$recommender[[2]]
Recommender of type 'IBCF' for 'realRatingMatrix'
learned using 4000 users.
$recommender[[3]]
```

```
Recommender of type 'UBCF' for 'realRatingMatrix'
learned using 4000 users.
$recommender[[4]]
Recommender of type 'RANDOM' for 'realRatingMatrix'
learned using 4000 users.
$weights
[1] 0.2 0.3 0.3 0.2
```

Observe the weights assignment in the hybrid model. We see that the popular and random recommenders are assigned 20% weight each, whereas the ITCF and UBCF methods involved in the preceding hybrid are assigned 30% weight each. It is not mandatory to set the weights while building a hybrid recommender, in which case, equal weights are assigned to each of the methods involved in the hybrid recommender. Now that our model is ready, let's make predictions and evaluate the performance with the following code:

```
# making predictions
pred <- predict(hybrid_recom, getData(Jester5k_es, "known"),
type="ratings")
# # set the predictions that fall outside the valid range to the boundary
values
pred@data@x[pred@data@x[] < -10] <- -10
pred@data@x[pred@data@x[] > 10] <- 10
# calculating performance measurements
hybrid_recom_pred = calcPredictionAccuracy(pred, getData(Jester5k_es,
"unknown"))
# printing the performance measurements
library(knitr)
print(kable(hybrid_recom_pred))
```

This will result in the following output:

```
|      |        x|
|:-----|--------:|
|RMSE  | 4.468849|
|MSE   | 19.970611|
|MAE   | 3.493577|
```

Summary

In this chapter, we used the recommenderlab library extensively to build the various types of joke-recommendation engines based on the Jester jokes dataset. We also learned about the theoretical concepts behind the methods.

Recommender systems is an individual ML area on its own. This subject is so vast that it cannot be covered in just one chapter. Several types of recommendation systems exists and they may be applied to datasets in specific scenarios. Matrix factorization, singular-value decomposition approximation, most popular items, and SlopeOne are some techniques that may be employed to build recommendation systems. These techniques are outside the scope of this chapter as these are rarely used in business situations to build recommendation systems, and the aim of the chapter is provide exposure to more popular techniques. Further learning on recommendation engines could be in the direction of exploring and studying these rarely-used techniques and applying them to real-world problems.

The next chapter is focused on NLP techniques. We are going to implement a sentiment-analysis engine on Amazon product reviews using several popular techniques. We'll explore semantic and syntactic approaches to analyzing text and then apply them on the Amazon review corpus. I am all geared up to turn this page and move on to the next chapter. How about you?!

References

While the `recommenderlab` library is super popular in the R community, this is not the only choice for building a recommendation system. Here are some other popular libraries you may rely on to implement recommendation engines:

- `rrecsys`: There are several popular recommendation systems, such as Global/Item/User-Average baselines, Item-Based KNN, FunkSVD, BPR, and weighted ALS for rapid prototyping. Refer to `https://cran.r-project.org/web/packages/rrecsys/index.htmlImplementations` for more information.

- `recosystem`: The R wrapper of the `libmf` library (`http://www.csie.ntu.edu.tw/~cjlin/libmf/`) for recommender system using matrix factorization. It is typically used to approximate an incomplete matrix using the product of two matrices in a latent space. Other common names for this task include collaborative filtering, matrix completion, and matrix recovery. High-performance multicore parallel computing is supported in this package.

- `rectools`: An advanced package for recommender systems to incorporate user and item covariate information, including item category preferences with parallel computation, novel variations on statistical latent factor model, focus group finder, NMF, ANOVA, and cosine models.

4

Sentiment Analysis of Amazon Reviews with NLP

Every day, we generate data from emails, online posts such as blogs, social media comments, and more. It is not surprising to say that unstructured text data is much larger in size than the tabular data that exists in the databases of any organization. It is important for organizations to acquire useful insights from the text data pertaining to the organization. Due to the different nature of the text data when compared to data in databases, the methods that need to be employed to understand the text data are different. In this chapter, we will learn a number of key techniques in **natural language processing** (**NLP**) that help us to work on text data.

The common definition of NLP is as follows: an area of computer science and artificial intelligence that deals with the interactions between computers and human (natural) languages; in particular, how to program computers to fruitfully process large amounts of natural language data.

In general terms, NLP deals with understanding human speech as it is spoken. It helps machines read and understand "text".

Human languages are highly complex and several ambiguities need to be resolved in order to correctly comprehend the spoken language or written text. In the area of NLP, several techniques are applied in order to deal with these ambiguities, including the **Part-of-Speech** (**POS**) tagger, term disambiguation, entity extraction, relations' extraction, key term recognition, and more.

For natural language systems to work successfully, a consistent knowledge base, such as a detailed thesaurus, a lexicon of words, a dataset for linguistic and grammatical rules, an ontology, and up-to-date entities, are prerequisites.

It may be noted that NLP is concerned with understanding the text from not just the syntactic perspective, but also from a semantic perspective. Similar to humans, the idea is for the machines to be able to perceive underlying messages behind the spoken words and not just the structure of words in sentences. There are numerous application areas of NLP, and the following are just a few of these:

- Speech recognition systems
- Question answering systems
- Machine translation
- Text summarization
- Virtual agents or chatbots
- Text classification
- Topic segmentation

As the NLP subject area in itself is very vast, it is not practical to cover all the areas in just one chapter. Therefore, we will be focusing on "text classification" in this chapter. We do this by implementing a project that performs sentiment analysis in the reviews expressed by Amazon.com customers. Sentiment analysis is a type of text classification task where we classify each of the documents (reviews) into one of the possible categories. The possible categories could be positive, negative, or neutral, or it could be positive, negative, or a rating on a scale of 1 to 10.

Text documents that need to be classified cannot be input directly to a machine learning algorithm. Each of the documents needs to be represented in a certain format that is acceptable for the ML algorithm as input to work on. In this chapter, we explore, implement, and understand the **Bag of Words** (**BoW**) word embedding approaches. These are approaches in which text can be represented.

As we progress with the chapter, we will cover the following topics:

- The sentiment analysis problem
- Understanding the Amazon reviews dataset
- Building a text sentiment classifier with the BoW approach
- Understanding word embedding approaches
- Building a text sentiment classifier with pretrained Word2Vec word embedding based on Reuters news corpus
- Building a text sentiment classifier with GloVe word embedding
- Building a text sentiment classifier with fastText

The sentiment analysis problem

Sentiment analysis is one of the most general text classification applications. The purpose of it is to analyze messages such as user reviews, and feedback from employees, in order to identify whether the underlying sentiment is positive, negative, or neutral.

Analyzing and reporting sentiment in texts allows businesses to quickly get a consolidated high-level insight without having to read each one of the comments received.

While it is possible to generate holistic sentiment based on the overall comments received, there is also an extended area called **aspect-based sentiment analysis**. It is focused on deriving sentiment based on each area of the service. For example, a customer that visited a restaurant when writing a review would generally cover areas such as ambience, food quality, service quality, and price. Though the feedback about each of the areas may not be quoted under a specific heading, the sentences in the review comments would naturally cover the customer's opinion of one or more of these areas. Aspect-based sentiment analysis attempts to identify the sentences in the reviews in each of the areas and then identify whether the sentiment is positive, negative, or neutral. Providing sentiment by each area helps businesses quickly identify their weak areas.

In this chapter, we will discuss and implement methods that are aimed at identifying the overall sentiment from the review texts. The task can be achieved in several ways, ranging from a simple lexicon method to a complex word embedding method.

A **lexicon** method is not really a machine learning method. It is more a rule based method that is based on a predefined positive and negative words dictionary. The method involves looking up the number of positive words and negative words in each review. If the count of positive words in the review is more than the count of negative words, then the review is marked as positive, otherwise it is marked as negative. If there are an equal number of positive and negative words, then the review is marked as neutral. As implementing this method is straightforward, and as it comes with a requirement for a predefined dictionary, we will not cover the implementation of the lexicon method in this chapter.

While it is possible to consider the sentiment analysis problem as an unsupervised clustering problem, in this chapter we consider it as a supervised classification problem. This is because, we have the Amazon reviews labeled dataset available. We can make use of these labels to build classification models, and therefore, the supervised algorithm.

Getting started

The dataset is available for download and use at the following URL:

```
https://drive.google.com/drive/u/0/folders/0Bz8a_
Dbh9Qhbfll6bVpmNUtUcFdjYmF2SEpmZUZUcVNiMUw1TWN6RDV3a0JHT3kxLVhVR2M.
```

Understanding the Amazon reviews dataset

We use the Amazon product reviews polarity dataset for the various projects in this chapter. It is an open dataset constructed and made available by Xiang Zhang. It is used as a text classification benchmark in the paper: *Character-level Convolutional Networks for Text Classification* and *Advances in Neural Information Processing Systems* 28, *Xiang Zhang, Junbo Zhao, Yann LeCun, (NIPS 2015)*.

The Amazon reviews polarity dataset is constructed by taking review score 1 and 2 as negative, 4 and 5 as positive. Samples of score 3 are ignored. In the dataset, class 1 is the negative and class 2 is the positive. The dataset has 1,800,000 training samples and 200,000 testing samples.

The `train.csv` and `test.csv` files contains all the samples as comma-separated values. There are three columns in them, corresponding to class index (1 or 2), review title, and review text. The review title and text are escaped using double quotes ("), and any internal double quote is escaped by 2 double quotes (""). New lines are escaped by a backslash followed with an "n" character that is "\n".

To ensure that we are able to run our projects, even with minimal infrastructure, let's restrict the number of records to be considered in our dataset to 1,000 records only. Of course, the code that we use in the projects can be extended to any number of records, as long as the hardware infrastructure support is available. Let's first read the data and visualize the records with the following code:

```
# reading first 1000 reviews
reviews_text<-readLines('/home/sunil/Desktop/sentiment_analysis/amazon
_reviews_polarity.csv', n = 1000)
# converting the reviews_text character vector to a dataframe
reviews_text<-data.frame(reviews_text)
# visualizing the dataframe
View(reviews_text)
```

This will result in the following output:

	reviews_text
1	"2","Stuning even for the non-gamer","This sound tr...
2	"2","The best soundtrack ever to anything.","I'm rea...
3	"2","Amazing!","This soundtrack is my favorite musi...
4	"2","Excellent Soundtrack","I truly like this soundtra...
5	"2","Remember, Pull Your Jaw Off The Floor After He...
6	"2","an absolute masterpiece","I am quite sure any ...
7	"1","Buyer beware","This is a self-published book, a...

Showing 1 to 8 of 1,000 entries

Post reading the file, we can see that there is only one column in the dataset and this column had both the review text and the sentiment components in it. We will slightly modify the format of the dataset for the purpose of using it with sentiment analysis projects in this chapter involving the BoW, Word2vec, and GloVe approaches. Let's modify the format of the dataset with the following code:

```
# separating the sentiment and the review text
# post separation the first column will have the first 4 characters
# second column will have the rest of the characters
# first column should be named "Sentiment"
# second column to be named "SentimentText"
library(tidyr)
reviews_text<-separate(data = reviews_text, col = reviews_text, into =
c("Sentiment", "SentimentText"), sep = 4)
# viewing the dataset post the column split
View(reviews_text)
```

This will result in the following output:

	Sentiment	SentimentText
1	"2",	"Stuning even for the non-gamer","This sound track...
2	"2",	"The best soundtrack ever to anything.","I'm readin...
3	"2",	"Amazing!","This soundtrack is my favorite music of...
4	"2",	"Excellent Soundtrack","I truly like this soundtrack a...
5	"2",	"Remember, Pull Your Jaw Off The Floor After Hearin...
6	"2",	"an absolute masterpiece","I am quite sure any of y...
7	"1",	"Buyer beware","This is a self-published book, and if...

Showing 1 to 8 of 1,000 entries

Now we have two columns in our dataset. However, there is unnecessary punctuation that exists in both the columns that may cause problems with processing the dataset further. Let's attempt to remove the punctuation with the following code:

```
# Retaining only alphanumeric values in the sentiment column
reviews_text$Sentiment<-gsub("[^[:alnum:] ]","",reviews_text$Sentiment)
# Retaining only alphanumeric values in the sentiment text
reviews_text$SentimentText<-gsub("[^[:alnum:] ]","
",reviews_text$SentimentText)
# Replacing multiple spaces in the text with single space
reviews_text$SentimentText<-gsub("(?<=[\\s])\\s*|^\\s+|\\s+$", "",
reviews_text$SentimentText, perl=TRUE)
# Viewing the dataset
View(reviews_text)
# Writing the output to a file that can be consumed in other projects
write.table(reviews_text,file =
"/home/sunil/Desktop/sentiment_analysis/Sentiment Analysis
Dataset.csv",row.names = F,col.names = T,sep=',')
```

This will result in the following output:

	Sentiment	SentimentText
1	2	Stuning even for the non gamer This sound track w...
2	2	The best soundtrack ever to anything I m reading a ...
3	2	Amazing This soundtrack is my favorite music of all ...
4	2	Excellent Soundtrack I truly like this soundtrack and...
5	2	Remember Pull Your Jaw Off The Floor After Hearing ...
6	2	an absolute masterpiece I am quite sure any of you ...
7	1	Buyer beware This is a self published book and if yo...

Showing 1 to 8 of 1,000 entries

From the preceding output, we see that we have a clean dataset that is ready for use. Also, we have written the output to a file. When we build the sentiment analyzer, we can start directly reading the dataset from the Sentiment Analysis Dataset.csv file.

The fastText algorithm expects the dataset to be in a different format. The data input to fastText should comply the following format:

```
__label__<X>   <Text>
```

In this example, X is the class name. Text is the actual review text that led to the rating specified under the class. Both the rating and text should be placed on one line with no quotes. The classes are __label__1 and __label__2, and there should be only one class per row. Let's accomplish the fastText library required format with the following code block:

```
# reading the first 1000 reviews from the dataset
reviews_text<-readLines('/home/sunil/Desktop/sentiment_analysis/amazon
_reviews_polarity.csv', n = 1000)
# basic EDA to confirm that the data is read correctly
print(class(reviews_text))
print(length(reviews_text))
print(head(reviews_text,2))
# replacing the positive sentiment value 2 with __label__2
reviews_text<-gsub("\\\"2\\\",","__label__2 ",reviews_text)
# replacing the negative sentiment value 1 with __label__1
reviews_text<-gsub("\\\"1\\\",","__label__1 ",reviews_text)
# removing the unnecessary \" characters
reviews_text<-gsub("\\\""," ",reviews_text)
# replacing multiple spaces in the text with single space
reviews_text<-gsub("(?<=[\\s])\\s*|^\\s+|\\s+$", "", reviews_text,
```

```
perl=TRUE)
# Basic EDA post the required processing to confirm input is as desired
print("EDA POST PROCESSING")
print(class(reviews_text))
print(length(reviews_text))
print(head(reviews_text,2))
# writing the revamped file to the directory so we could use it with
# fastText sentiment analyzer project
fileConn<-file("/home/sunil/Desktop/sentiment_analysis/Sentiment Analysis
Dataset_ft.txt")
writeLines(reviews_text, fileConn)
close(fileConn)
```

This will result in the following output:

```
[1] "EDA PRIOR TO PROCESSING"
[1] "character"
[1] 1000
[1] "\"2\",\"Stuning even for the non-gamer\",\"This sound track was
beautiful! It paints the senery in your mind so well I would recomend it
even to people who hate vid. game music! I have played the game Chrono
Cross but out of all of the games I have ever played it has the best music!
It backs away from crude keyboarding and takes a fresher step with grate
guitars and soulful orchestras. It would impress anyone who cares to
listen! ^_^\""
[2] "\"2\",\"The best soundtrack ever to anything.\",\"I'm reading a lot of
reviews saying that this is the best 'game soundtrack' and I figured that
I'd write a review to disagree a bit. This in my opinino is Yasunori
Mitsuda's ultimate masterpiece. The music is timeless and I'm been
listening to it for years now and its beauty simply refuses to fade.The
price tag on this is pretty staggering I must say, but if you are going to
buy any cd for this much money, this is the only one that I feel would be
worth every penny.\""
[1] "EDA POST PROCESSING"
[1] "character"
[1] 1000\
[1] "__label__2 Stuning even for the non-gamer , This sound track was
beautiful! It paints the senery in your mind so well I would recommend it
even to people who hate vid. game music! I have played the game Chrono
Cross but out of all of the games I have ever played it has the best music!
It backs away from crude keyboarding and takes a fresher step with grate
guitars and soulful orchestras. It would impress anyone who cares to
listen! ^_^"
[2] "__label__2 The best soundtrack ever to anything. , I'm reading a lot
of reviews saying that this is the best 'game soundtrack' and I figured
that I'd write a review to disagree a bit. This in my opinino is Yasunori
Mitsuda's ultimate masterpiece. The music is timeless and I'm been
listening to it for years now and its beauty simply refuses to fade. The
```

price tag on this is pretty staggering I must say, but if you are going to buy any cd for this much money, this is the only one that I feel would be worth every penny."

From the output of basic EDA code, we can see that the dataset is in the required format, therefore we can proceed to our next section of implementing the sentiment analysis engine using the BoW approach. Along side the implementation, we will delve into learning the concept behind the approach, and explore the sub-techniques that can be leveraged in the approach to obtain better results.

Building a text sentiment classifier with the BoW approach

The intent of the BoW approach is to convert the review text provided into a matrix form. It represents documents as a set of distinct words by ignoring the order and meaning of the words. Each row of the matrix represents each review (otherwise called a document in NLP), and the columns represent the universal set of words present in all the reviews. For each document, and across each word, the existence of the word, or the frequency of the word occurrence, in that specific document is recorded. Finally, the matrix created from word frequency vectors represents the documents set. This methodology is used to create input datasets that are required to train the models, and also to prepare the test dataset that need to be used by the trained models to perform text classification. Now that we understand the BoW motivation, let's jump into implementing the steps to build a sentiment analysis classifier based on this approach, as shown in the following code block:

```
# including the required libraries
library(SnowballC)
library(tm)
# setting the working directory where the text reviews dataset is located
# recollect that we pre-processed and transformed the raw dataset format
setwd('/home/sunil/Desktop/sentiment_analysis/')
# reading the transformed file as a dataframe
text <- read.table(file='Sentiment Analysis Dataset.csv', sep=',',header =
TRUE)
# checking the dataframe to confirm everything is in tact
print(dim(text))
View(text)
```

This will result in the following output:

```
> print(dim(text))
[1] 1000 2
> View(text)
```

	Sentiment	SentimentText
1	2	Stuning even for the non gamer This sound track w...
2	2	The best soundtrack ever to anything I m reading a ...
3	2	Amazing This soundtrack is my favorite music of all ...
4	2	Excellent Soundtrack I truly like this soundtrack and...
5	2	Remember Pull Your Jaw Off The Floor After Hearing ...
6	2	an absolute masterpiece I am quite sure any of you ...
7	1	Buyer beware This is a self published book and if yo...
8	2	Glorious story I loved Whisper of the wicked saints T...

Showing 1 to 8 of 1,000 entries

The first step in processing text data involves creating a *corpus*, which is a collection of text documents. The VCorpus function in the tm package enables conversion of the reviews comments column in the data frame into a volatile corpus. This can be achieved through the following code:

```
# transforming the text into volatile corpus
train_corp = VCorpus(VectorSource(text$SentimentText))
print(train_corp)
```

This will result in the following output:

```
> print(train_corp)
<<VCorpus>>
Metadata:  corpus specific: 0, document level (indexed): 0
Content:  documents: 1000
```

From the volatile corpus, we create a **Document Term Matrix (DTM)**. A DTM is a sparse matrix that is created using the tm library's DocumentTermMatrix function. The rows of the matrix indicate documents and the columns indicate features, that is, words. The matrix is sparse because all unique unigram sets of the dataset become columns in DTM and, as each review comment does not have all elements of the unigram set, most cells will have a 0, indicating the absence of the unigram.

While it is possible to extract n-grams (unigrams, bigrams, trigrams, and so on) as part of the BoW approach, the tokenize parameter can be set and passed as part of the control list in the `DocumentTermMatrix` function to accomplish n-grams in DTM. It must be noted that using n-grams as part of the DTM creates a very high number of columns in the DTM. This is one of the demerits of the BoW approach, and, in some cases, it could stall the execution of the project due to limited memory. As our specific case is also limited by hardware infrastructure, we restrict ourselves by including only the unigrams in DTM in this project. Apart from just generating unigrams, we also perform some additional processing on the reviews text document by passing parameters to the control list in the `tm` library's `DocumentTermMatrix` function. The processing we do on the review text documents during the creation of the DTM is given here:

1. Change the case of the text to lowercase.
2. Remove any numbers.
3. Remove stop words using the English language stop word list from the Snowball stemmer project. Stop words are common words, such as a, an, in, and the, that do not add value in deciding sentiment based on review comments.
4. Remove punctuation.
5. Perform stemming, which aims at resolving a word into the base form of the word, that is, strip the plural *s* from nouns, the *ing* from verbs, or other affixes. A stem is a natural group of words with equal or very similar meaning. After the stemming process, every word is represented by its stem. The `SnowballC` library provides the capability to obtain the root for each of the words in the review comments.

Let's now create a DTM from the volatile corpus and do the text preprocessing with the following code block:

```
# creating document term matrix
dtm_train <- DocumentTermMatrix(train_corp, control = list(
  tolower = TRUE, removeNumbers = TRUE,
  stopwords = TRUE,
  removePunctuation = TRUE,
  stemming = TRUE
))
# Basic EDA on dtm
inspect(dtm_train)
```

This will result in the following output:

```
> inspect(dtm_train)
<<DocumentTermMatrix (documents: 1000, terms: 5794)>>
Non-/sparse entries: 34494/5759506
```

```
Sparsity             : 99%
Maximal term length: 21
Weighting            : term frequency (tf)
Sample               :
     Terms
Docs  book can get great just like love one read time
 111    0    3   2     0    0    0    2   1    0    2
 162    4    1   0     0    0    1    0   0    1    0
 190    0    0   0     0    0    0    0   0    0    0
 230    0    1   1     0    0    0    1   0    0    0
 304    0    0   0     0    0    3    0   2    0    0
 399    0    0   0     0    0    0    0   0    0    0
 431    9    1   0     0    0    1    2   0    0    1
 456    1    0   0     0    0    0    0   1    2    0
 618    0    2   3     1    4    1    3   1    0    1
  72    0    0   1     0    2    0    0   1    0    1
```

We see from the output that there are 1,000 documents that were processed and form rows in the matrix. There are 5,794 columns representing unique unigrams from the reviews following the additional text processing. We also see that the DTM is 99% sparse and consists of non-zero entries only in 34,494 cells. The non-zero cells represent the frequency of occurrence of the word on the column in the document represent on the row of the DTM. The weighting is done through the default 'term frequency' weighting, as we did not specify any weighting parameter in the control list supplied to the `DocumentTermMatrix` function. Other forms of weighting, such as **term frequency-inverse document frequency (TFIDF)**, are also possible just by passing the appropriate weight parameter in the control list to the `DocumentTermMatrix` function. For now, we will stick to weighting based on term frequency, which is the default. We also see from the `inspect` function that several sample documents were output along with the term frequencies in these documents.

The DTM tends to get very big, even for normal sized datasets. Removing sparse terms, that is, terms occurring only in very few documents, is the technique that can be tried to reduce the size of the matrix without losing significant relations inherent to the matrix. Let's remove sparse columns from the matrix. We will attempt to remove those terms that have at least a 99% of sparse elements with the following line of code:

```
# Removing sparse terms
dtm_train= removeSparseTerms(dtm_train, 0.99)
inspect(dtm_train)
```

This will result in the following output:

```
> inspect(dtm_train)
<<DocumentTermMatrix (documents: 1000, terms: 686)>>
Non-/sparse entries: 23204/662796
Sparsity             : 97%
```

```
Maximal term length: 10
Weighting          : term frequency (tf)
Sample             :
      Terms
Docs  book can get great just like love one read time
  174    0   0   1     1    1    2    0   2    0    1
  304    0   0   0     0    0    3    0   2    0    0
  355    3   0   0     0    1    1    2   3    1    0
  380    4   1   0     0    1    0    0   1    0    2
  465    5   0   1     1    0    0    0   2    6    0
  618    0   2   3     1    4    1    3   1    0    1
   72    0   0   1     0    2    0    0   1    0    1
  836    1   0   0     0    0    3    0   0    5    1
  866    8   0   1     0    0    1    0   0    4    0
  959    0   0   2     1    1    0    0   2    0    1
```

We now see from the output of the `inspect` function that the sparsity of the matrix is reduced to 97%, and the number of unigrams (columns of the matrix) is reduced to 686. We are now ready with the DTM that can be used for training with any machine learning classification algorithm. In the next few lines of code, let's attempt to divide our DTM into training and test dataset:

```
# splitting the train and test DTM
dtm_train_train <- dtm_train[1:800, ]
dtm_train_test <- dtm_train[801:1000, ]
dtm_train_train_labels <- as.factor(as.character(text[1:800, ]$Sentiment))
dtm_train_test_labels <- as.factor(as.character(text[801:1000,
]$Sentiment))
```

We will be using a machine learning algorithm called **Naive Bayes** to create a model. Naive Bayes is generally trained on data with nominal features. We can observe that the cells in our DTM are numeric and therefore need to be converted to nominal prior to feeding the dataset as input for creating the model with Naive Bayes. As each cell indicates the word frequency in the review, and as the number of times a word used in the review does not impact sentiment, let's write a function to convert the cell values with a non-zero value to Y, and in case of a zero, let's convert it to N, with the following code:

```
cellconvert<- function(x) {
x <- ifelse(x > 0, "Y", "N")
}
```

Now, let's apply the function on all rows of the training dataset, and test dataset we have previously created in this project with the following code:

```
# applying the function to rows in training and test datasets
dtm_train_train <- apply(dtm_train_train, MARGIN = 2,cellconvert)
```

```
dtm_train_test <- apply(dtm_train_test, MARGIN = 2,cellconvert)
# inspecting the train dtm to confirm all is in tact
View(dtm_train_train)
```

This will result in the following output:

	abl	absolut	act	action	actual	adam	add	adult	adventur	advertis	age	ago	agre
1	N	N	N	N	N	N	N	N	N	N	N	N	N
2	N	N	N	N	N	N	N	N	N	N	N	N	N
3	N	Y	N	N	N	N	N	N	N	N	N	N	N
4	N	N	N	N	N	N	N	N	N	N	N	N	N
5	N	N	N	N	N	N	N	N	N	N	N	N	N
6	N	Y	N	N	Y	N	N	N	N	N	N	N	N
7	N	N	N	N	N	N	N	N	N	N	N	N	N

Showing 1 to 8 of 800 entries

We can see from the output that all the cells in the training and test DTMs are now converted to nominal values. Thus, let's proceed to build a text sentiment analysis classifier using the Naive Bayes algorithm from the e1071 library, as follows:

```
# training the naive bayes classifier on the training dtm
library(e1071)
nb_senti_classifier=naiveBayes(dtm_train_train,dtm_train_train_labels)
# printing the summary of the model created
summary(nb_senti_classifier)
```

This will result in the following output:

```
> summary(nb_senti_classifier)
        Length Class  Mode
apriori 2      table  numeric
tables  686    -none- list
levels  2      -none- character
call    3      -none- call
```

The preceding summary output shows that the nb_senti_classifier object is successfully created from the training DTM. Let's now use the model object to predict sentiment on the test data DTM. In the following code block, we are instructing that the predictions should be classes and not prediction probabilities:

```
# making predictions on the test data dtm
nb_predicts<-predict(nb_senti_classifier, dtm_train_test,type="class")
# printing the predictions from the model
print(nb_predicts)
```

This will result in the following output:

```
[1] 1 1 2 1 1 1 1 1 1 2 2 1 2 2 2 2 1 2 1 1 2 1 2 1 1 1 2 2 1 2 2 2 2 1 2 1
1 1 1 2 2 2 2 1 2 1 1 1 1 1 1 1 1 1 1 1 2 1 1 1 2 1 1 1 1 1 1 2 1 1 2 2
1 2 2 2 2 1 2 2 1 1 1 1 1 2 1 1 2 1 1 1 1 1 2 2 2 2 2 1 2 2 1 2 1 1 1 1 2
2 2 2 2 1 1 1 2 2 2 1 1 1 1 1 2 1 2 1 1 1 1 1 1 1 1 2 1 1 1 1 1 2 1 1 1
1 1 2 1 1 1 1 1 1 2 2 2 2 1 2 2 1 2 2 1 1 2 2 1 1 2 2 2 2 2 2 2 2 2 2 1
1 2 1 2 1 2 2 1 1 1 1 2
Levels: 1 2
```

With the following code, let us now compute the accuracy of the model using the `mmetric` function in the `rminer` library:

```
# computing accuracy of the model
library(rminer)
print(mmetric(nb_predicts, dtm_train_test_labels, c("ACC")))
```

This will result in the following output:

```
[1] 79
```

We achieved a 79% accuracy just with a very quick and basic BoW model. The model can be further improved by means of techniques such as parameter tuning, lemmatization, new features creation, and so on.

Pros and cons of the BoW approach

Now that we have an understanding of both the theory and implementation of the BoW approach, let's examine the pros and cons of the approach. When it comes to pros, the BoW approach is very simple to understand and implement and therefore offers a lot of flexibility for customization on any text dataset. It may be observed that the BoW approach does not retain the order of words specifically when only unigrams are considered. This problem is generally overcome by retaining n-grams in the DTM. However, it comes at the cost as larger infrastructure is needed to process the text and build a classifier. Another severe drawback of the approach is that it does not respect the semantics of the word. For example, the words "car" and "automobile" are often used in the same context. A model built based on BoW treats the sentences "buy used cars" and "purchase old automobiles" as very different sentences. While these sentences are the same, BoW models do not classify these sentences as the same, as the words in these sentences are not matching. It is possible to consider the semantics of words in sentences using an approach called word embedding. This is something we will explore in our next section.

Understanding word embedding

The BoW models that we discussed in our earlier section suffer from a problem that they do not capture information about a word's meaning or context. This means that potential relationships, such as contextual closeness, are not captured across collections of words. For example, the approach cannot capture simple relationships, such as determining that the words "cars" and "buses" both refer to vehicles that are often discussed in the context of transportation. This problem that we experience with the BoW approach will be overcome by word embedding, which is an improved approach to mapping semantically similar words.

Word vectors represent words as multidimensional continuous floating point numbers, where semantically similar words are mapped to proximate points in geometric space. For example, the words *fruit* and *leaves* would have a similar word vector, *tree*. This is due to the similarity of their meanings, whereas the word *television* would be quite distant in the geometrical space. In other words, words that are used in a similar context will be mapped to a proximate vector space.

The word vectors can be of n dimensions, and n can take any number as input from the user creating it (for example 10, 70, 500). The dimensions are latent in the sense that it may not be apparent to humans what each of these dimensions means in reality. There are methods such as **Continuous Bag of Words** (**CBOW**) and **Skip-Gram** that enable conceiving the word vectors from the text provided as training input to word embedding algorithms. Also, the individual numbers in the word vector represent the word's distributed weight across dimensions. In a general sense, each dimension represents a latent meaning, and the word's numerical weight on that dimension captures the closeness of its association with and to that meaning. Thus, the semantics of the word are embedded across the dimensions of the vector.

Though the word vectors are multidimensional and cannot be visualized directly, it is possible to visualize the vectors learned, by projecting them down to two dimensions using techniques such as the t-SNE dimensionality reduction technique. The following diagram displays learned word vectors in two dimensional spaces for country capitals, verb tenses, and gender relationships:

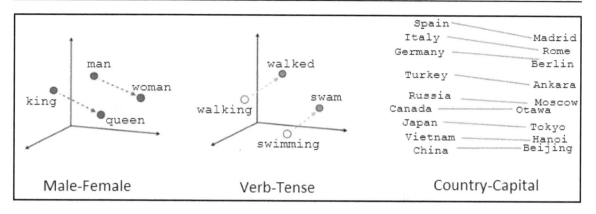

Visualization of word embeddings in a two dimensional space

When we observe the word embedding visualization, we can perceive that the vectors captured some general, and in fact quite useful, semantic information about words and their relationships to one another. With this, each word in the text now can be represented as a row in the matrix similar to that of the BoW approach, but, unlike the BoW approach, it captures the relationships between the words.

The advantage of representing words as vectors is that they lend themselves to mathematical operators. For example, we can add and subtract vectors. The canonical example here is showing that by using word vectors we can determine the following:

$$king - man + woman = queen$$

In the given example, we subtracted the gender (man) from the word vector for king and added another gender (woman), and we obtained a new word vector from the operation (*king - man + woman*) that maps most closely to the word vector for queen.

A few more amazing examples of mathematical operations that can be achieved on word vectors are shown as follows:

- Given two words, we can establish the degree of similarity between them:

  ```
  model.similarity('woman','man')
  ```

  And the output is as follows:

  ```
  0.73723527
  ```

- Finding the odd one out from the set of words given as input:

```
model.doesnt_match('breakfast cereal dinner lunch';.split())
```

The odd one is given as the following output:

```
'cereal'
```

- Derive analogies, for example:

```
model.most_similar(positive=['woman','king'],negative=['man'],topn=
1)
```

The output is as follows:

```
queen: 0.508
```

Now, what it all means for us is that machines are able to identify semantically similar words given in a sentence. The following diagram is a gag related to word embedding that made me laugh, but the gag does convey the power of word embedding application, which otherwise would not be possible with the BoW kind of text representations:

A gag demonstrating the power of word embeddings application

There are several techniques that can be used to learn word embedding from text data. Word2vec, GloVe, and fastText are some of the popular techniques. Each of these techniques allows us to either train our own word embedding from the text data we have, or use the readily available pretrained vectors.

This approach of learning our own word embedding requires a lot of training data and can be slow, but this option will learn an embedding both targeted to the specific text data and the NLP task at hand.

Pretrained word embedding vectors are vectors that are trained on large amounts of text data (usually billions of words) available on sources such as Wikipedia. These are generally high-quality word embedding vectors made available by companies such as Google or Facebook. We can download these pretrained vector files and consume them to obtain word vectors for the words in the text that we would like to classify or cluster.

Building a text sentiment classifier with pretrained word2vec word embedding based on Reuters news corpus

Word2vec was developed by Tomas Mikolov, et al. at Google in 2013 as a response to making the neural-network-based training of the embedding more efficient, and since then it has become the de facto standard for developing pretrained word embedding.

Word2vec introduced the following two different learning models to learn the word embedding:

- **CBOW**: Learns the embedding by predicting the current word based on its context.
- **Continuous Skip-Gram**: The continuous Skip-Gram model learns by predicting the surrounding words given a current word.

Both CBOW and Skip-Gram methods of learning are focused on learning the words given their local usage context, where the context of the word itself is defined by a window of neighboring words. This window is a configurable parameter of the model.

The `softmaxreg` library in R offers pretrained `word2vec` word embedding that can be used for building our sentiment analysis engine for the Amazon reviews data. The pretrained vector is built using the `word2vec` model, and it is based on the `Reuter_50_50` dataset, UCI Machine Learning Repository (`https://archive.ics.uci.edu/ml/datasets/ Reuter_50_50`).

Without any delay, let's get into the code and also review the approach followed in this code:

```
# including the required library
library(softmaxreg)
# importing the word2vec pretrained vector into memory
data(word2vec)
```

Let's examine the `word2vec` pretrained emdeddings. It is just another data frame, and therefore can be reviewed through the regular `dim` and `View` commands as follows:

```
View(word2vec)
```

This will result in the following output:

	word	col1	col2	col3	col4	col5	col6	col7	col8	col9	col10	col11	col12
1	expands	0.40553	0.28853	-0.59429	-0.15864	0.17330	0.22890	0.12649	-0.07011	0.19290	-0.18450	-0.31630	-0.0274
2	mobutu	0.00938	0.41061	-0.32526	0.08943	0.24560	-0.07285	0.15669	-0.20358	0.17647	-0.42533	-0.23850	-0.0907
3	contends	0.08888	0.25632	-0.35454	0.09304	-0.04802	0.40575	-0.14148	-0.36319	0.06299	-0.09494	-0.04483	-0.0474
4	june	2.60395	2.67175	0.79562	2.66639	-0.05200	-4.05117	-0.56082	3.75234	-0.25007	-3.44570	1.47910	-4.3133
5	branch	0.07003	-0.23289	-0.60203	-0.10870	-0.19456	0.44149	0.71168	-0.06072	0.58567	-0.57049	-0.29232	0.6115
6	boom	0.74027	0.42692	-0.68575	-0.37356	-0.04878	-0.82011	0.24457	-0.24025	0.30356	0.13888	-0.19215	0.1878
7	book	1.33032	-0.41911	-0.36447	0.43531	-0.43499	-0.08896	0.06012	-0.40035	1.04810	-0.49407	-0.29468	0.0840

Showing 1 to 8 of 12,853 entries

Here, let's use the following `dim` command:

```
dim(word2vec)
```

This will result in the following output:

```
[1] 12853 21
```

From the preceding output, we can observe that there are `12853` words that have got word vectors in the pretrained vector. Each of the words is defined using 20 dimensions, and these dimensions define the context of the words. In the next step, we can look up the word vector for each of the words in the review comments. As there are only 12,853 words in the pretrained word embedding, there is a possibility that we encounter a word that does not exist in the pretrained embedding. In such a case, the unidentified word is represented with a 20 dimension vector that is filled with zeros.

We also need to understand that the word vectors are available only at a word level, and therefore in order to decode the entire review, we take the mean of all the word vectors of the words that made up the review. Let's review the concept of getting the word vector for a sentence from individual word vectors with an example.

Assume the sentence we want to get the word vector for is, *it is very bright and sunny this morning*. Individual words that comprise the sentence are *it, is, very, bright, and, sunny, this,* and *morning*.

Now, we can look up each of these words in the pretrained vector and get the corresponding word vectors as shown in the following table:

Word	dim1	dim2	dim3	.....	....	dim19	dim20
it	-2.25	0.75	1.75	-1.25	-0.25	-3.25	-2.25
is	0.75	1.75	1.75	-2.25	-2.25	0.75	-0.25
very	-2.25	2.75	1.75	-0.25	0.75	0.75	-2.25
bright	-3.25	-3.25	-2.25	-1.25	0.75	1.75	-0.25
and	-0.25	-1.25	-2.25	2.75	-3.25	-0.25	1.75
sunny	0	0	0	0	0	0	0
this	-2.25	-3.25	2.75	0.75	-0.25	-0.25	-0.25
morning	-0.25	-3.25	-2.25	1.75	0.75	2.75	2.75

Now, we have word vectors that comprise the sentence. Please note that these are not actual word vector values but just are made up to demonstrate the approach. Also, observe that the word `sunny` is represented with zeros across the dimensions to symbolize that the word is not found in the pretrained word embedding. In order to get the word vector for the sentence, we just compute the mean of each dimension. The resulting vector is a 1 x 20 vector representing the sentence, as follows:

Sentence	-1.21875	-0.71875	0.15625	0.03125	-0.46875	0.28125	-0.09375

The `softmaxreg` library offers the `wordEmbed` function where we could pass a sentence and ask it to compute the `mean` word vector for the sentence. The following code is a custom function that was created to apply the `wordEmbed` function on each of the Amazon reviews we have in hand. At the end of applying this function to the reviews dataset, we expect to have a *n* x 20 matrix that is the word vector representation of our reviews. The *n* in the *n* x 20 represents the number of rows and 20 is the number of dimensions through which each review is represented, as seen in the following code:

```
# function to get word vector for each review
docVectors = function(x)
{
   wordEmbed(x, word2vec, meanVec = TRUE)
}
# setting the working directory and reading the reviews dataset
setwd('/home/sunil/Desktop/sentiment_analysis/')
text = read.csv(file='Sentiment Analysis Dataset.csv', header = TRUE)
# applying the docVector function on each of the reviews
# storing the matrix of word vectors as temp
temp=t(sapply(text$SentimentText, docVectors))
# visualizing the word vectors output
View(temp)
```

This will result in the following output:

1	0.9001672	0.0141028750	-1.0616862	-0.04416625	-0.3533675	-0.6738171	0.0729040000	-0.8279434	-0.45585687	0.3550463
2	1.1277574	-0.1360358824	-1.2177180	-0.15301882	-0.9068942	-1.1244064	-0.3733900980	-0.9224640	-0.18435745	0.3772624
3	0.8885871	-0.0856260294	-1.1432905	-0.01933257	-0.7176180	-0.7664569	0.0773944853	-0.6180129	-0.17120868	0.2428186
4	0.5495152	-0.2092989344	-1.1647043	-0.35540713	-0.5236916	-0.3633348	0.3810807377	-0.3947897	-0.05829189	0.0426281
5	1.0133256	-0.0268535556	-1.1575648	0.03562800	-0.3391036	-0.6837794	-0.0377413333	-0.5346906	-0.43420989	-0.1457108
6	0.8082624	-0.2846747297	-1.2788924	-0.15989108	-0.7584445	-0.6270216	0.0809281757	-0.5134580	-0.17634953	0.0652181
7	0.8902767	0.0157431081	-1.2022697	-0.14477669	-0.9319141	-0.7504433	-0.1275757432	-0.7498420	-0.38869155	0.3180145

Showing 1 to 8 of 1.000 entries

Then we review `temp` using the `dim` command, as follows:

```
dim(temp)
```

This will result in the following output:

```
1000 20
```

We can see from the output that we have word vectors created for each of the reviews in our corpus. This data frame can now be used to build classification models using an ML algorithm. The following code to achieve classification is no different from the one we did for the BoW approach:

```
# splitting the dataset into train and test
temp_train=temp[1:800,]
temp_test=temp[801:1000,]
labels_train=as.factor(as.character(text[1:800,]$Sentiment))
labels_test=as.factor(as.character(text[801:1000,]$Sentiment))
# including the random forest library
library(randomForest)
# training a model using random forest classifier with training dataset
# observe that we are using 20 trees to create the model
rf_senti_classifier=randomForest(temp_train, labels_train,ntree=20)
print(rf_senti_classifier)
```

This will result in the following output:

```
randomForest(x = temp_train, y = labels_train, ntree = 20)
               Type of random forest: classification
                     Number of trees: 20
No. of variables tried at each split: 4
        OOB estimate of  error rate: 44.25%
Confusion matrix:
      1    2 class.error
1 238 172    0.4195122
2 182 208    0.4666667
```

The preceding output shows that the Random Forest model object is successfully created. Of course, the model can be improved further; however we are not going to be doing that here as the focus is to demonstrate making use of word embeddings, rather than getting the best performing classifier.

Next, with the following code we make use of the Random Forest model to make predictions on the test data and then report out the performance:

```
# making predictions on the dataset
rf_predicts<-predict(rf_senti_classifier, temp_test)
library(rminer)
print(mmetric(rf_predicts, labels_test, c("ACC")))
```

This will result in the following output:

```
[1] 62.5
```

We see that we get a 62% accuracy from using the pretrained `word2vec` embeddings made out of the Reuters news group's dataset.

Building a text sentiment classifier with GloVe word embedding

Stanford University's Pennington, et al. developed an extension of the `word2vec` method that is called **Global Vectors for Word Representation (GloVe)** for efficiently learning word vectors.

GloVe combines the global statistics of matrix factorization techniques, such as LSA, with the local context-based learning in `word2vec`. Also, unlike `word2vec`, rather than using a window to define local context, GloVe constructs an explicit word context or word co-occurrence matrix using statistics across the whole text corpus. As an effect, the learning model yields generally better word embeddings.

The `text2vec` library in R has a GloVe implementation that we could use to train to obtain word embeddings from our own training corpus. Alternatively, pretrained GloVe word embeddings can be downloaded and reused, similar to the way we did in the earlier `word2vec` pretrained embedding project covered in the previous section.

The following code block demonstrates the way in which GloVe word embeddings can be created and used for sentiment analysis, or, for that matter, any text classification task. We are not going to discuss explicitly the steps involved, since the code is already heavily commented with detailed explanations of each of the steps:

```
# including the required library
library(text2vec)
# setting the working directory
setwd('/home/sunil/Desktop/sentiment_analysis/')
# reading the dataset
text = read.csv(file='Sentiment Analysis Dataset.csv', header = TRUE)
# subsetting only the review text so as to create Glove word embedding
wiki = as.character(text$SentimentText)
# Create iterator over tokens
tokens = space_tokenizer(wiki)
# Create vocabulary. Terms will be unigrams (simple words).
it = itoken(tokens, progressbar = FALSE)
```

```
vocab = create_vocabulary(it)
# consider a term in the vocabulary if and only if the term has appeared
aleast three times in the dataset
vocab = prune_vocabulary(vocab, term_count_min = 3L)
# Use the filtered vocabulary
vectorizer = vocab_vectorizer(vocab)
# use window of 5 for context words and create a term co-occurance matrix
tcm = create_tcm(it, vectorizer, skip_grams_window = 5L)
# create the glove embedding for each each in the vocab and
# the dimension of the word embedding should set to 50
# x_max is the maximum number of co-occurrences to use in the weighting
# function
# note that training the word embedding is time consuming - be patient
glove = GlobalVectors$new(word_vectors_size = 50, vocabulary = vocab, x_max
= 100)
wv_main = glove$fit_transform(tcm, n_iter = 10, convergence_tol = 0.01)
```

This will result in the following output:

```
INFO [2018-10-30 06:58:14] 2018-10-30 06:58:14 - epoch 1, expected cost
0.0231
INFO [2018-10-30 06:58:15] 2018-10-30 06:58:15 - epoch 2, expected cost
0.0139
INFO [2018-10-30 06:58:15] 2018-10-30 06:58:15 - epoch 3, expected cost
0.0114
INFO [2018-10-30 06:58:15] 2018-10-30 06:58:15 - epoch 4, expected cost
0.0100
INFO [2018-10-30 06:58:15] 2018-10-30 06:58:15 - epoch 5, expected cost
0.0091
INFO [2018-10-30 06:58:15] 2018-10-30 06:58:15 - epoch 6, expected cost
0.0084
INFO [2018-10-30 06:58:16] 2018-10-30 06:58:16 - epoch 7, expected cost
0.0079
INFO [2018-10-30 06:58:16] 2018-10-30 06:58:16 - epoch 8, expected cost
0.0074
INFO [2018-10-30 06:58:16] 2018-10-30 06:58:16 - epoch 9, expected cost
0.0071
INFO [2018-10-30 06:58:16] 2018-10-30 06:58:16 - epoch 10, expected cost
0.0068
```

The following uses the `glove` model to obtain the combined word vector:

```
# Glove model learns two sets of word vectors - main and context.
# both matrices may be added to get the combined word vector
wv_context = glove$components
word_vectors = wv_main + t(wv_context)
# converting the word_vector to a dataframe for visualization
word_vectors=data.frame(word_vectors)
```

```
# the word for each embedding is set as row name by default
# using the tibble library rownames_to_column function, the rownames is
copied as first column of the dataframe
# we also name the first column of the dataframe as words
library(tibble)
word_vectors=rownames_to_column(word_vectors, var = "words")
View(word_vectors)
```

This will result in the following output:

	words	X1	X2	X3	X4	X5	X6	X7	X8
1	proper	-0.183938861	0.535530344	-0.27412909	-0.461746097	-0.100504830	-0.06466068	0.27132151	0.022755206
2	Stone	-0.217368635	-0.619701624	0.10591805	0.089803070	-0.080577239	0.22282777	-0.08934084	0.214210983
3	practically	-0.178972661	-0.437024154	0.33109763	0.293728501	-0.078952595	0.30375123	0.02330199	-0.322362080
4	sings	-0.724405587	-0.028475652	0.49158170	-0.093503183	-0.710965484	0.18689734	0.08935627	0.252694234
5	Lit	0.062119633	0.026841171	0.47210883	0.196634409	0.005582331	0.72392994	0.24716580	-0.867472053
6	Jordan	0.904864848	-0.067597866	0.54857185	-0.549472168	-0.369129092	-0.86961806	-0.44172353	-0.154189382
7	Waist	0.568695709	-0.044349857	0.40858194	0.849112123	0.691710413	-0.78834027	-0.47516816	0.633008987
8	planet	-0.387996703	-0.714768857	0.34405428	0.011786014	0.091809094	0.16621399	-0.70615846	0.718172520
9	replacement	-0.072499663	-0.221350074	-0.55283985	0.209103354	0.317845784	-0.27957728	0.05697817	-0.054497272

Showing 1 to 10 of 3,019 entries

We make use of the `softmaxreg` library to obtain the mean word vector for each review. This is similar to what we did in `word2vec` pretrained embedding in the previous section. Observe that we are passing our own trained word embedding `word_vectors` to the `wordEmbed()` function, as follows:

```
library(softmaxreg)
docVectors = function(x)
{
   wordEmbed(x, word_vectors, meanVec = TRUE)
}
# applying the function docVectors function on the entire reviews dataset
# this will result in word embedding representation of the entire reviews #
dataset
temp=t(sapply(text$SentimentText, docVectors))
View(temp)
```

This will result in the following output:

1	-1.021674e-01	0.22374366	-0.31380372	-0.24523438	-0.2180971	-0.21172789	0.186605388	0.20956048	-0.04558712	-0.051
2	1.194950e-02	0.14968572	-0.30805228	-0.18293664	-0.2258637	-0.25498353	0.132375430	0.20017790	-0.10048077	-0.187
3	-4.014470e-02	0.21875645	-0.35748125	-0.21297610	-0.2207336	-0.22556214	0.183079215	0.16168858	-0.04248742	-0.077
4	-4.899985e-02	0.29007017	-0.32021912	-0.18001094	-0.1704606	-0.16176222	0.190220987	0.17813243	-0.05338900	0.081
5	-2.130447e-02	0.28866528	-0.37526279	-0.19773100	-0.2535055	-0.19100561	0.109212284	0.16018515	-0.07803226	0.011
6	1.814156e-02	0.24571046	-0.36655681	-0.16840549	-0.2477891	-0.14538835	0.142705234	0.19516599	-0.01567144	-0.100
7	-4.428909e-03	0.18590056	-0.30330482	-0.13953408	-0.1885616	-0.19064436	0.114466673	0.21589199	-0.07120269	-0.192
8	1.132957e-01	0.18632012	-0.37151759	-0.18653199	-0.2790091	-0.24254169	0.170006885	0.37763887	-0.11926454	-0.229
9	6.110208e-02	0.19142639	-0.30230704	-0.21159849	-0.2664249	-0.26331480	0.172434261	0.20212321	-0.08137470	-0.149
10	6.803773e-04	0.20381966	0.35296762	0.30145897	0.2980702	0.27664309	0.127662254	0.28347248	0.11008705	0.207

Showing 1 to 11 of 1,000 entries

We will now split the dataset into train and test portions, and use the `randomforest` library to build a model to train, as shown in the following lines of code:

```
# splitting the dataset into train and test portions
temp_train=temp[1:800,]
temp_test=temp[801:1000,]
labels_train=as.factor(as.character(text[1:800,]$Sentiment))
labels_test=as.factor(as.character(text[801:1000,]$Sentiment))
# using randomforest to build a model on train data
library(randomForest)
rf_senti_classifier=randomForest(temp_train, labels_train,ntree=20)
print(rf_senti_classifier)
```

This will result in the following output:

```
Call:
 randomForest(x = temp_train, y = labels_train, ntree = 20)
               Type of random forest: classification
                     Number of trees: 20
No. of variables tried at each split: 7


        OOB estimate of  error rate: 42.12%
Confusion matrix:
     1    2 class.error
1 250 160   0.3902439
2 177 213   0.4538462
```

Then, we use the Random Forest model created to predict labels, as follows:

```
# predicting labels using the randomforest model created
rf_predicts<-predict(rf_senti_classifier, temp_test)
# estimating the accuracy from the predictions
library(rminer)
print(mmetric(rf_predicts, labels_test, c("ACC")))
```

This will result in the following output:

```
[1] 66.5
```

With this method, we obtain an accuracy of 66%. This is despite the fact that the word embeddings are obtained from words in just 1,000 text samples. The model may be further improved by using a pretrained embedding. The overall framework to use the pretrained embedding remains the same as what we did in `word2vec` project in the previous section.

Building a text sentiment classifier with fastText

`fastText` is a library and is an extension of `word2vec` for word representation. It was created by the Facebook Research Team in 2016. While Word2vec and GloVe approaches treat words as the smallest unit to train on, fastText breaks words into several n-grams, that is, subwords. For example, the trigrams for the word apple are app, ppl, and ple. The word embedding for the word apple is sum of all the word n-grams. Due to the nature of the algorithm's embedding generation, fastText is more resource-intensive and takes additional time to train. Some of the advantages of `fastText` are as follows:

- It generates better word embeddings for rare words (including misspelled words).
- For out of vocabulary words, fastText can construct the vector for a word from its character n-grams, even if a word doesn't appear in training corpus. This is not a possibility for both Word2vec and GloVe.

The `fastTextR` library provides an interface to the fastText. Let's make use of the `fastTextR` library for our project to build a sentiment analysis engine on Amazon reviews. While it is possible to download pretrained fastText word embedding and make use of it for our project, let's make an attempt to train a word embedding based on the reviews dataset we have in hand. It should be noted that the approach in terms of making use of fastText pretrained word embedding is similar to the approach we followed in the `word2vec` based project that we dealt with earlier.

Similar to the project covered in the previous section, comments are included inline in the code. The comments explain each of the lines indicating the approach taken to build the Amazon reviews sentiment analyzer in this project. Let's look into the following code now:

```
# loading the required libary
library(fastTextR)
# setting the working directory
setwd('/home/sunil/Desktop/sentiment_analysis/')
# reading the input reviews file
# recollect that fastText needs the file in a specific format and we
created one compatiable file in
# "Understanding the Amazon Reviews Dataset" section of this chaptertext =
readLines("Sentiment Analysis Dataset_ft.txt")
# Viewing the text vector for conformation
View(text)
```

This will result in the following output:

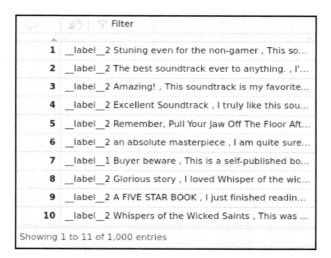

Now let's divide the reviews into training and test datasets, and view them using the following lines of code:

```
# dividing the reviews into training and test
temp_train=text[1:800]temp_test=text[801:1000]
# Viewing the train datasets for confirmation
View(temp_train)
```

This will give the following output:

	V1
1	__label__2 Stuning even for the non-gamer , This sound track was beautiful! It paints the senery in your mind so well I would recomend it eve
2	__label__2 The best soundtrack ever to anything. , I'm reading a lot of reviews saying that this is the best 'game soundtrack' and I figured that
3	__label__2 Amazing! , This soundtrack is my favorite music of all time, hands down. The intense sadness of Prisoners of Fate (which means all
4	__label__2 Excellent Soundtrack , I truly like this soundtrack and I enjoy video game music. I have played this game and most of the music on
5	__label__2 Remember, Pull Your Jaw Off The Floor After Hearing it , If you've played the game, you know how divine the music is! Every single
6	__label__2 an absolute masterpiece , I am quite sure any of you actually taking the time to read this have played the game at least once, and
7	__label__1 Buyer beware , This is a self-published book, and if you want to know why--read a few paragraphs! Those 5 star reviews must have
8	__label__2 Glorious story , I loved Whisper of the wicked saints. The story was amazing and I was pleasantly surprised at the changes in the bo
9	__label__2 A FIVE STAR BOOK , I just finished reading Whisper of the Wicked saints. I fell in love with the caracters. I expected an average rom
10	__label__2 Whispers of the Wicked Saints , This was a easy to read book that made me want to keep reading on and on, not easy to put down
11	__label__1 The Worst! , A complete waste of time. Typographical errors, poor grammar, and a totally pathetic plot add up to absolutely nothin
12	__label__2 Great book , This was a great book,I just could not put it down,and could not read it fast enough. Boy what a book the twist and tu
13	__label__2 Great Read , I thought this book was brilliant, but yet realistic. It showed me that to error is human. I loved the fact that this writer s
14	__label__1 Oh please , I guess you have to be a romance novel lover for this one, and not a very discerning one. All others beware! It is absolu
15	__label__1 Awful beyond belief! , I feel I have to write to keep others from wasting their money. This book seems to have been written by a 7t
16	__label__1 Don't try to fool us with fake reviews. , It's glaringly obvious that all of the glowing reviews have been written by the same person,
17	__label__2 A romantic zen baseball comedy , When you hear folks say that they don't make 'em like that anymore, they might be talking abou
18	__label__2 Fashionable Compression Stockings! , After I had a DVT my doctor required me to wear compression stockings. I wore ugly white T
19	__label__2 Jobst UltraSheer Thigh High , Excellent product. However, they are very difficult to get on for older people. I feel like I've had a full
20	__label__1 sizes recomended in the size chart are not real , sizes are much smaller than what is recomended in the chart. I tried to put it and s
21	__label__1 mens ultrasheer , This model may be ok for sedentary types, but I'm active and get around alot in my job - consistently found these
22	__label__2 Delicious cookie mix , I thought it was funny that I bought this product without knowing it was a mix. I read the header very quickly
23	__label__1 Another Abysmal Digital Copy , Rather than scratches and insect droppings, this one has random pixelations combined with mudd
24	__label__2 A fascinating insight into the life of modern Japanese teens , I thoroughly enjoyed Rising Sons and Daughters. I don't know of any
25	__label__2 i liked this album more then i thought i would , I heard a song or two and thought same o same o,but when i listened to songs like
26	__label__1 Problem with charging smaller AAAs , I have had the charger for more than two years. It charges AA batteries just fine, but has a h
27	__label__1 Works, but not as advertised , I bought one of these chargers..the instructions say the lights stay on while the battery charges...true
28	__label__1 Disappointed , I read the reviews,made my purchase and was very disappointed. The charger is convenient by charging all four bat
29	__label__1 Oh dear , I was excited to find a book ostensibly about Muslim feminism, but this volume did not live up to the expectations.One e

Use the following code to view the test dataset:

```
View(temp_test)
```

This will give the following output:

We will now create a `.txt` file for the train and test dataset using the following code:

```
# creating txt file for train and test dataset
# the fasttext function expects files to be passed for training and testing
fileConn<-file("/home/sunil/Desktop/sentiment_analysis/train.ft.txt")
writeLines(temp_train, fileConn)
close(fileConn)
fileConn<-file("/home/sunil/Desktop/sentiment_analysis/test.ft.txt")
writeLines(temp_test, fileConn)
close(fileConn)
# creating a test file with no labels
# recollect the original test dataset has labels in it
# as the dataset is just a subset obtained from full dataset
temp_test_nolabel<- gsub("__label__1", "", temp_test, perl=TRUE)
temp_test_nolabel<- gsub("__label__2", "", temp_test_nolabel, perl=TRUE)
```

Now we will view the no labels test dataset for confirmation using the following command:

```
View(temp_test_nolabel)
```

This will result in the following output:

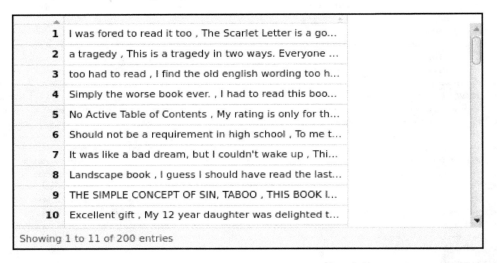

Let's now write the no labels test dataset to a file so we can use it for testing, as follows:

```
fileConn<-
file("/home/sunil/Desktop/sentiment_analysis/test_nolabel.ft.txt")
writeLines(temp_test_nolabel, fileConn)
close(fileConn)
# training a supervised classification model with training dataset file
model<-fasttext("/home/sunil/Desktop/sentiment_analysis/train.ft.txt",
method = "supervised", control = ft.control(nthreads = 3L))
# Obtain all the words from a previously trained model=
words<-get_words(model)
# viewing the words for confirmation. These are the set of words present  #
in our training data
View(words)
```

This will result in the following output:

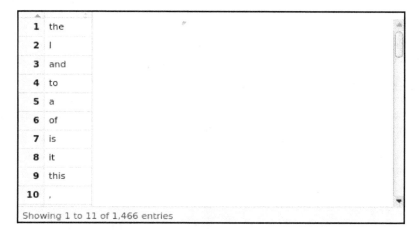

Now we will obtain the word vectors from a previously trained model and view the word vectors for each word in our training dataset, as follows:

```
# Obtain word vectors from a previously trained model.
word_vec<-get_word_vectors(model, words)
# Viewing the word vectors for each word in our training dataset
# observe that the word embedding dimension is 5
View(word_vec)
```

This will result in the following output:

the	0.0048942002	0.0119173910	-0.007131079	-0.0538859000	0.0631141519
I	-0.0063322317	0.0340317730	0.013798622	0.1236231849	0.1272250041
and	0.0792929615	-0.0941958507	-0.025565661	0.0298055482	0.0359003235
to	0.0081054736	-0.0231800877	0.072222504	-0.0636691093	-0.0123805664
a	0.0155201275	0.0465942640	0.082956703	-0.0899057742	-0.0420212336
of	0.0261684284	0.0131066605	-0.099206429	0.0289594168	0.0892733419
is	-0.1270621065	0.0093145892	-0.113851033	-0.0712698530	0.0564273112
it	-0.0264658609	-0.0369169824	-0.133076520	-0.1075496329	0.0285819056
this	0.0263026568	0.0273612377	-0.044421507	0.0267953683	-0.0022563544
,	0.0069336370	-0.1357625574	-0.036102463	-0.1234983876	-0.0015986823

Showing 1 to 11 of 1,466 entries

We will predict the labels for the reviews in the no labels test dataset and write it to a file for future reference. Then we will get the predictions into a data frame to compute the performance and see the estimate of the accuracy using the following lines of code:

```
# predicting the labels for the reviews in the no labels test dataset
# and writing it to a file for future reference
predict(model, newdata_file=
"/home/sunil/Desktop/sentiment_analysis/test_nolabel.ft.txt",result_file="/
home/sunil/Desktop/sentiment_analysis/fasttext_result.txt")
# getting the predictions into a dataframe so as to compute performance   #
measurementft_preds<-predict(model, newdata_file=
"/home/sunil/Desktop/sentiment_analysis/test_nolabel.ft.txt")
# reading the test file to extract the actual labels
reviewstestfile<
readLines("/home/sunil/Desktop/sentiment_analysis/test.ft.txt")
# extracting just the labels frm each line
library(stringi)
actlabels<-stri_extract_first(reviewstestfile, regex="\\w+")
# converting the actual labels and predicted labels into factors
actlabels<-as.factor(as.character(actlabels))
ft_preds<-as.factor(as.character(ft_preds))
# getting the estimate of the accuracy
library(rminer)
print(mmetric(actlabels, ft_preds, c("ACC")))
```

This will result in the following output:

```
[1] 58
```

We have a 58% accuracy with the `fastText` method on our reviews data. As a next step, we could check whether the accuracy may be further improved by making use of fastText pretrained word embedding. As we already know, implementing a project by making use of pretrained embedding is not very different from the implementation that we followed in the `word2vec` project described in the earlier section of this chapter. The difference is just that the training step to obtain word embedding needs to be discarded and the model variable in the code covered in this project code should be initiated with the pretrained word embeddings.

Summary

In this chapter, we learned various NLP techniques, namely BoW, Word2vec, GloVe, and fastText. We built projects involving these techniques to perform sentiment analysis on an Amazon reviews dataset. The projects that were built involved two approaches, making use of pretrained word embeddings and building the word embeddings from our own dataset. We tried both these approaches to represent text in a format that can be consumed by ML algorithms that resulted in models with the ability to perform sentiment analysis.

In the next chapter, we will learn about customer segmentation by making use of a wholesale dataset. We will look at customer segmentation as an unsupervised problem and build projects with various techniques that can identify inherent groups within the e-commerce company's customer base. Come, let's explore the world of building an e-commerce customer segmentation engine with ML!

Summary

5
Customer Segmentation Using Wholesale Data

In today's competitive world, the success of an organization largely depends on how much it understands its customers' behavior. Understanding each customer individually to better tailor the organizational effort to individual needs is a very expensive task. Based on the size of the organization, this task can be very challenging as well. As an alternative, organizations rely on something called **segmentation**, which attempts to categorize customers into groups based on identified similarities. This critical aspect of customer segmentation allows organizations to extend their efforts to the individual needs of various customer subsets (if not catering to individual needs), therefore reaping greater benefits.

In this chapter, we will learn about the concept and importance of customer segmentation. We'll then deep dive into learning the various **machine learning** (**ML**) methods to identify subgroups of customers based on customer characteristics. We'll implement several projects using the wholesale dataset to understand the ML techniques for segmentation. In the next section, we'll start by learning the foundations of customer segmentation and the need for ML techniques to achieve segmentation. We will cover the following topics as we progress:

- Understanding customer segmentation
- Understanding the wholesale customer dataset and the segmentation problem
- Identifying the customer segments in wholesale customer data using DIANA
- Identifying the customer segments in wholesale customer data using AGNES

Understanding customer segmentation

Customer segmentation, or market segmentation, at a basic level, is the partitioning of a broad range of potential customers in a given market into specific subgroups of customers, where each of the subgroups contains customers that share certain similarities. The following diagram depicts the formal definition of customer segmentation where customers are identified into three groups:

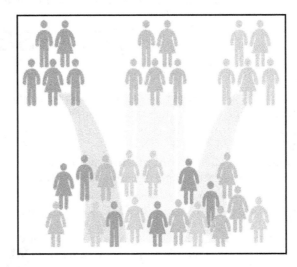

Illustration depicting customer segmentation definition

Customer segmentation needs the organizations to gather data about customers and analyze it to identify patterns that can be used to determine subgroups. The segmentation of customers could be achieved through multiple data points related to customers. The following are some of the data points:

- **Demographics**: This data point includes race, ethnicity, age, gender, religion, level of education, income, life stage, marital status, occupation
- **Psychographics**: This data point includes lifestyle, values, socioeconomic standing, personality
- **Behavioral**: This data point includes product usage, loyalties, awareness, occasions, knowledge, liking, and purchase patterns

With billions of people in the world, efficiently making use of customer segmentation will help organizations narrow down the pool and reach only the people that mean something to their business, ultimately driving conversions and revenue. The following are some of the specific objectives that organizations attempt to achieve through identifying segments in their customers:

- Identifying higher-percentage opportunities that the sales team can pursue
- Identifying customer groups that have a higher interest in the product, and customize the product according to the needs of high-interest customers
- Developing very focused marketing messages to specific customer groups so as to drive higher-quality inbound interest in the product
- Choosing the best communication channel for various segments, which might be email, social media, radio, or another approach, depending on the segment
- Concentrating on the most profitable customers
- Upselling and cross-selling other products and services
- Test pricing options
- Identifying new product or service opportunities

When an organization needs to perform segmentation, it can typically look for common characteristics, such as shared needs, common interests, similar lifestyles, or even similar demographic profiles and come up with segments in customer data. Unfortunately, creating segments is not that simple. With the availability of big data, organizations now have hundreds of characteristics of customers they can look at in order to come up with segments. It is not feasible for a person or few people in an organization to go through hundreds of types of data, find relationships between each of them, and then establish segments based on several different values possible for each data point. That's where unsupervised ML, called **clustering,** comes to rescue.

Clustering is the mechanism of using ML algorithms to identify relationships in different types of data, thereby yielding new segments based on those relationships. Simply put, clustering finds the relationship between data points so they can be segmented.

The terms **cluster analysis** and **customer segmentation** are closely related and used interchangeably by ML practitioners. However, there is an important difference between these terms.

Clustering is a tool that helps organizations put together data based on similarities and statistical connections. Clustering is very helpful in guiding the development of suitable customer segments. It also provides useful statistical measures of the potential target customers. While the objective for an organization is to identify effective customer segments from data, simply applying a clustering technique on data and grouping the data in itself may or may not offer effective customer segments. This essentially means that the output obtained from clustering, that is, the **clusters**, need to be further analyzed to get insight into the meaning of each of the clusters, and then determine which clusters can be utilized for downstream activities, such as business promotions. The following is a flow diagram that helps us to understand the role of clustering in the customer-segmentation process:

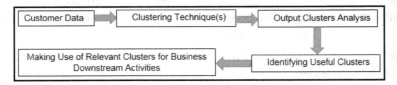

Role of clustering in customer segmentation

Now that we understand that clustering forms a stepping stone to performing customer segmentation, in the rest of the chapter, we will discuss various clustering techniques and implement projects around these techniques to create customer segments. For our projects, we make use of the wholesale customer dataset. Before delving into the projects, let's learn about the dataset and perform **exploratory data analysis** (**EDA**) to get a better understanding of the data.

Understanding the wholesale customer dataset and the segmentation problem

The UCI Machine Learning Repository offers the wholesale customer dataset at `https://archive.ics.uci.edu/ml/datasets/wholesale+customers`. The dataset refers to clients of a wholesale distributor. It includes the annual spending in **monetary units** (**m.u.**) on diverse product categories. The goal of these projects is to apply clustering techniques to identify segments that are relevant for certain business activities, such as rolling out a marketing campaign. Before we actually use the clustering algorithms to get clusters, let's first read the data and perform some EDA to understand the data using the following code block:

```
# setting the working directory to a folder where dataset is located
setwd('/home/sunil/Desktop/chapter5/')
# reading the dataset to cust_data dataframe
cust_data = read.csv(file='Wholesale_customers_ data.csv', header = TRUE)
# knowing the dimensions of the dataframe
print(dim(cust_data))
Output :
440 8
# printing the data structure
print(str(cust_data))
'data.frame': 440 obs. of 8 variables:
 $ Channel : int 2 2 2 1 2 2 2 2 1 2 ...
 $ Region : int 3 3 3 3 3 3 3 3 3 3 ...
 $ Fresh : int 12669 7057 6353 13265 22615 9413 12126 7579...
 $ Milk : int 9656 9810 8808 1196 5410 8259 3199 4956...
 $ Grocery : int 7561 9568 7684 4221 7198 5126 6975 9426...
 $ Frozen : int 214 1762 2405 6404 3915 666 480 1669...
 $ Detergents_Paper: int 2674 3293 3516 507 1777 1795 3140 3321...
 $ Delicassen : int 1338 1776 7844 1788 5185 1451 545 2566...
# Viewing the data to get an intuition of the data
View(cust_data)
```

This will give the following output:

	Channel	Region	Fresh	Milk	Grocery	Frozen	Detergents_Paper	Delicassen
1	2	3	12669	9656	7561	214	2674	1338
2	2	3	7057	9810	9568	1762	3293	1776
3	2	3	6353	8808	7684	2405	3516	7844
4	1	3	13265	1196	4221	6404	507	1788
5	2	3	22615	5410	7198	3915	1777	5185
6	2	3	9413	8259	5126	666	1795	1451
7	2	3	12126	3199	6975	480	3140	545
8	2	3	7579	4956	9426	1669	3321	2566
9	1	3	5963	3648	6192	425	1716	750
10	2	3	6006	11093	18881	1159	7425	2098

Showing 1 to 10 of 440 entries

Now let's check whether there are any entries with missing fields in our dataset:

```
# checking if there are any NAs in data
print(apply(cust_data, 2, function (x) sum(is.na(x))))
Output :
Channel Region Fresh Milk
0 0 0 0
Grocery Frozen Detergents_Paper Delicassen
```

```
0 0 0 0
# printing the summary of the dataset
print(summary(cust_data))
```

This will give the following output:

```
Channel Region Fresh Milk
  Min. :1.000 Min. :1.000 Min. : 3 Min. : 55
  1st Qu.:1.000 1st Qu.:2.000 1st Qu.: 3128 1st Qu.: 1533
  Median :1.000 Median :3.000 Median : 8504 Median : 3627
  Mean :1.323 Mean :2.543 Mean : 12000 Mean : 5796
  3rd Qu.:2.000 3rd Qu.:3.000 3rd Qu.: 16934 3rd Qu.: 7190
  Max. :2.000 Max. :3.000 Max. :112151 Max. :73498
  Grocery Frozen Detergents_Paper Delicassen
  Min. : 3.0 Min. : 3.0 Min. : 3 Min. : 25.0
  1st Qu.: 256.8 1st Qu.: 408.2 1st Qu.: 2153 1st Qu.: 742.2
  Median : 816.5 Median : 965.5 Median : 4756 Median : 1526.0
  Mean : 2881.5 Mean : 1524.9 Mean : 7951 Mean : 3071.9
  3rd Qu.: 3922.0 3rd Qu.: 1820.2 3rd Qu.:10656 3rd Qu.: 3554.2
  Max. :40827.0 Max. :47943.0 Max. :92780 Max. :60869.0
```

From the EDA, we see that there are `440` observations available in this dataset and there are eight variables. The dataset does not have any missing values. While the last six variables are goods that were brought by distributors from the wholesaler, the first two variables are factors (categorical variables) representing the location and channel of purchase. In our projects, we intend to identify the segments based on the sales into different products, therefore, the location and channel variables in the data are not very useful. Let's delete them from the dataset using the following code:

```
# excluding the non-useful columns from the dataset
cust_data<-cust_data[,c(-1,-2)]
# verifying the dataset post columns deletion
dim(cust_data)
```

This gives us the following output:

```
440 6
```

We see that only six columns are retained, confirming that the deletion of non-required columns is successful. From the summary output in the EDA code, we can also observe that the scale across all the retained columns is the same so we do not have to explicitly normalize the data.

It may be noted that most clustering algorithms involve computation of distance of some form (such as Euclidean, Manhattan, Grower). It is important that data is scaled across the columns in the dataset so as to ensure a variable does not end up as a dominating one in distance computation just because of high scale. In case of different scales observed in columns of the data, we will rely on techniques such as Z-transform or min-max transform. Applying one of these techniques on the data ensures that the columns of the dataset are scaled appropriately therefore leaving no dominating variables in the dataset to be used with clustering algorithms. Fortunately, we do not have this issue so we can continue with the dataset as it is.

Clustering algorithms impose identification of subgroups in the input dataset even if there are no clusters present. To ensure that we get meaningful clusters as output from the clustering algorithms, it is important to check whether clusters exist in the data at all. **Clustering tendency**, or the feasibility of the clustering analysis, is the process of identifying whether the clusters exist in the dataset. Given an input dataset, the process determines whether it has a non-random or non-uniform data structure distribution that will lead to meaningful clusters. The Hopkins statistic measure is used to determine cluster tendency. It takes a value between 0 and 1, and if the value of the Hopkins statistic is close to 0 (far below 0.5), it indicates the existence of valid clusters in the dataset. A Hopkins value closer to 1 indicates random structures in the dataset.

The `factoextra` library has a built-in `get_clust_tendency()` function that computes the Hopkins statistic on the input dataset. Let's apply this function on our wholesale dataset to determine whether the dataset is valid for clustering at all. The following code accomplishes the computation of the Hopkins statistic:

```
# setting the working directory to a folder where dataset is located
setwd('/home/sunil/Desktop/chapter5/')
# reading the dataset to cust_data dataframe
cust_data = read.csv(file='Wholesale_customers_ data.csv', header = TRUE)
# removing the non-required columns
cust_data<-cust_data[,c(-1,-2)]
# inlcuding the facto extra library
library(factoextra)
# computing and printing the hopikins statistic
print(get_clust_tendency(cust_data, graph=FALSE,n=50,seed = 123))
```

This will give the following output:

```
$hopkins_stat
[1] 0.06354846
```

The Hopkins statistic output for our dataset is very close to 0, so we can conclude that we have a dataset that is a good candidate for our clustering exercise.

Categories of clustering algorithms

There are numerous clustering algorithms available off the shelf in R. However, all these algorithms can be grouped into one of two categories:

- **Flat or partitioning algorithms**: These algorithms rely on an input parameter that defines the number of clusters to be identified in the dataset. The input parameter sometimes comes up as input from business directly or it can be established through certain statistical methods. For example, the **Elbow** method.
- **Hierarchical algorithms**: In these kinds of algorithms, the clusters are not identified in a single step. They involves multiple steps that run from a single cluster containing all the data points to *n* clusters containing single data point. Hierarchical algorithms can be further divided into the following two types:
  - **Divisive type**: A top-down clustering method where all points are initially assigned to a single cluster. In the next step, the cluster is split into two clusters which are least similar. The process of splitting the clusters is recursively done until each point has its own cluster, for example, the **DIvisive ANAlysis (DIANA)** clustering algorithm.
  - **Agglomerative type**: A bottom-up approach where, in the initial run, each point in the dataset is assigned *n* unique clusters, where *n* is equal to the number of observations in the dataset. In the next iteration, most similar clusters are merged (based on the distance between the clusters). The recursive process of merging the clusters continues until we are left with just one cluster, for example, **agglomerative nesting (AGNES)** algorithm.

As discussed earlier, there are numerous clustering algorithms available and we will focus on implementing projects using one algorithm for each type of clustering. We will implement project with k-means that is a flat or partitioning type clustering algorithm. We will then do customer segmentation with DIANA and AGNES, which are divisive and agglomerative, respectively.

Identifying the customer segments in wholesale customer data using k-means clustering

The k-means algorithm is perhaps the most popular and commonly-used clustering method from partitioning clustering type. Though we usually call the clustering algorithm k-means, multiple implementations of this algorithm exist, namely the **MacQueen, Lloyd and Forgy**, and **Hartigan-Wong** algorithms. It has been studied and found that the Hartigan-Wong algorithm performs better than the other two algorithms in most situations. K-means in R makes use of the Hartigan-Wong implementation by default.

The k-means algorithm requires the k-value to be passed as a parameter. The parameter indicates the number of clusters to be made with the input data. It is often a challenge for practitioners to determine the optimal k-value. Sometimes, we can go to a business and ask them how many clusters they would expect in the data. The answer from the business directly translates to be the *k* parameter value to be fed to the algorithm. In most cases though, the business is clueless as to the number of clusters. In such a case, the onus will be on the ML practitioner to determine the k-value. Fortunately, there are several methods available to determine this value. These methods can be classified into the following two categories:

- **Direct methods**: These methods rely on optimizing a criterion, such as *within cluster sums of squares* or *the average silhouette*. Examples of this method include the **V Elbow method** and the **V Silhouette method**.
- **Testing methods**: These methods consists of comparing evidence against a null hypothesis. Gap statistic is one popular example of this method.

In addition to Elbow, Silhouette, and gap statistic methods, there are more than 30 other indices and methods that have been published for identifying the optimal number of clusters. We will not delve into the theoretical details of these methods, as covering 30 methods in a single chapter is not practical. However, R offers an excellent library function, called NbClust that makes it easy for us to implement all these methods in one go. The NbClust function is so powerful that it determines the optimal clusters by varying all combinations of number of clusters, distance measures, and clustering methods and all in one go! Once the library function computes all 30 indices, the *majority rule* is applied on the output to determine the optimal number of clusters, that is, the k-value to be used as input to the algorithm. Let's implement NbClust for our wholesale dataset to determine the optimal k-value using the following code block:

```
# setting the working directory to a folder where dataset is located
setwd('/home/sunil/Desktop/chapter5/')
# reading the dataset to cust_data dataframe
cust_data = read.csv(file='Wholesale_customers_ data.csv', header = TRUE)
# removing the non-required columns
cust_data<-cust_data[,c(-1,-2)]
# including the NbClust library
library(NbClust)
# Computing the optimal number of clusters through the NbClust function
with distance as euclidean and using kmeans
NbClust(cust_data,distance="euclidean", method="kmeans")
```

This will give the following output:

```
*********************************************************************
* Among all indices:
* 1 proposed 2 as the best number of clusters
* 11 proposed 3 as the best number of clusters
* 2 proposed 4 as the best number of clusters
* 1 proposed 5 as the best number of clusters
* 4 proposed 8 as the best number of clusters
* 1 proposed 10 as the best number of clusters
* 1 proposed 12 as the best number of clusters
* 1 proposed 14 as the best number of clusters
* 1 proposed 15 as the best number of clusters
                    ***** Conclusion *****
* According to the majority rule, the best number of clusters is 3
*********************************************************************
```

As per the conclusion, we see the k-value that may be used for our problem is 3. Additionally, plotting an elbow curve with the total within-groups sums of squares against the number of clusters in a k-means solution can be helpful in determining the optimal number of clusters. K-means is defined by the objective function, which tries to minimize the sum of all squared distances within a cluster (intra-cluster distance) for all clusters. In the elbow-curve plotting method, we compute the intra-cluster distance with different values of k, and the intra-cluster distance with different k's is plotted as a graph. A bend in the elbow curve suggests the k-value that is optimal for the dataset. The elbow curve can be obtained within R using the following code block:

```
# computing the the intra-cluster distance with Ks ranging from 2 to 10
library(purrr)
tot_withinss <- map_dbl(2:10, function(k){
  model <- kmeans(cust_data, centers = k, nstart = 50)
  model$tot.withinss
})
# converting the Ks and computed intra-cluster distances to a dataframe
screeplot_df <- data.frame(k = 2:10,
```

```
                              tot_withinss = tot_withinss)
# plotting the elbow curve
library(ggplot2)
print( ggplot(screeplot_df, aes(x = k, y = tot_withinss)) +
        geom_line() +
        scale_x_continuous(breaks = 1:10) +
        labs(x = "k", y = "Within Cluster Sum of Squares") +
        ggtitle("Total Within Cluster Sum of Squares by # of Clusters
(k)") +
        geom_point(data = screeplot_df[2,], aes(x = k, y = tot_withinss),
              col = "red2", pch = 4, size = 7))
```

This will give the following output:

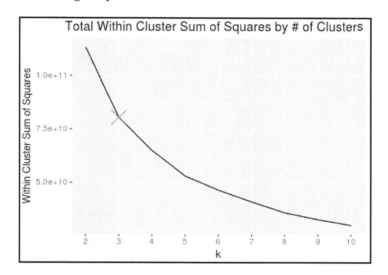

Even with the elbow curve method output, we see that the number of optimal clusters for our dataset is 3.

We see from the NbClust function that we have used the Euclidean distance as the distance. There are a number of distance types (euclidean, maximum, manhattan, canberra, binary, minkowski) that we could have used as values for this distance parameter in the NbClust function. Let's understand what this distance actually means. We are already aware that each observation in our dataset is formed by values that represent features. This essentially means each observation of our dataset can be represented as points in multidimensional space. If we have to say that two observations are similar, we would expect the distance between the two points in the multidimensional space to be lower, that is, both these points in multidimensional space are close to each other. A high distance value between the two points indicates that they are very dissimilar.

The Euclidean, Manhattan, and other types of distance measures are various ways in which distance can be measured given two points in a multidimensional space. Each of the distance measures involves a specific technique to compute the distance between the two points. The techniques involved in Manhattan and Euclidean, and the difference between their measures, are illustrated in the following screenshot:

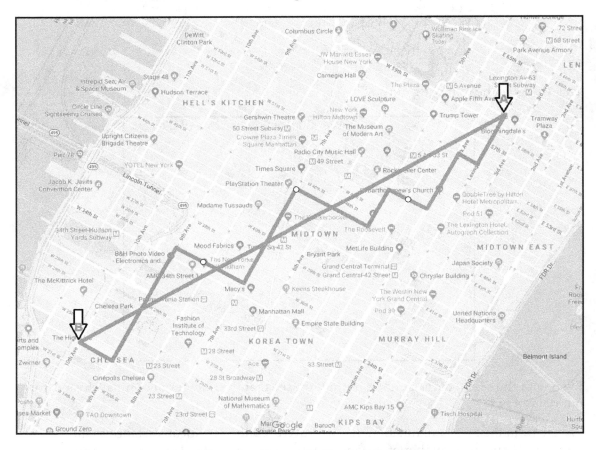

Difference between Manhattan and Euclidean distance measures

The Euclidean distance measures the shortest distance in the plane, whereas the Manhattan metric is the shortest path if one is allowed to move horizontally or vertically.

For example, if a and b are two points where a= (0,0), b = (3,4), then take a look at the following:

- dist_euclid (a,b) = sqrt(3^2+4^2) = 5
- dist_manhattan(a,b) = 3+4 = 7
- a=(a1,...,an), b=(b1,...,bn) (in *n* dimensions and points)
- dist_euclid (a,b) = sqrt((a1-b1)^2 + ... + (an-bn)^2)
- dist_manhattan(a,b) = sum(abs(a1-b1) + ... + abs(an-bn))

Both measure the shortest paths, but the Euclidean metric doesn't have any restrictions while the Manhattan metric only allows paths constant in all but one dimension.

Likewise, the other distance measures also involve a certain unique to measure the similarity between given points. We will not be going through each one of the techniques in detail in this chapter, but the idea to get is that a distance measure basically defines the level of similarity between given observations. It may be noted that a distance measure is not just used in NbClust but in multiple ML algorithms, including k-means.

Now that we've learned the various ways to identify our k-value and have implemented them to identify the optimal number of clusters for our wholesale dataset, let's implement the k-means algorithm with the following code:

```
library(cluster)
# runing kmeans in cust_data dataset to obtain 3 clusters
kmeansout <- kmeans(cust_data, centers = 3, nstart = 50)
print (kmeansout)
```

This will result in the following output:

```
> kmeansout
K-means clustering with 3 clusters of sizes 330, 50, 60
Cluster means:
      Fresh Milk Grocery Frozen Detergents_Paper Delicassen
1 8253.47 3824.603 5280.455 2572.661 1773.058 1137.497
2 8000.04 18511.420 27573.900 1996.680 12407.360 2252.020
3 35941.40 6044.450 6288.617 6713.967 1039.667 3049.467
Clustering vector:
  [1] 1 1 1 1 3 1 1 1 1 2 1 1 3 1 3 1 1 1 1 1 1 1 3 2 3 1 1 1 2 3 1 1 1 3 1
1 3 1 2 3 3 1 1 2 1 2 2 2 1 2 1 1 3 1
 [55] 3 1 2 1 1 1 1 2 1 1 1 2 1 1 1 1 1 1 1 1 1 1 1 2 1 1 1 1 1 1 1 2 2 3 1
3 1 1 2 1 1 1 1 1 1 1 1 1 3 1 1 1 1
[109] 1 2 1 2 1 1 1 1 1 1 1 1 1 1 1 1 3 3 1 1 1 3 1 1 1 1 1 1 1 1 1 1 1 3 3
1 1 2 1 1 1 3 1 1 1 1 1 2 1 1 1 1 1
[163] 1 2 1 2 1 1 1 1 1 2 1 2 1 1 3 1 1 1 3 1 3 1 1 1 1 1 1 1 1 1 1 1 1 1 3
1 1 1 2 2 3 1 1 2 1 1 1 2 1 2 1 1 1 1
```

```
[217] 2 1 1 1 1 1 1 1 1 1 1 1 1 1 1 1 3 1 1 1 1 1 1 3 3 3 1 1 1 1 1 1 1 1 1 1
2 1 3 1 3 1 1 3 3 1 1 3 1 1 2 2 1 2 1
[271] 1 1 1 3 1 1 3 1 1 1 1 1 3 3 3 3 1 1 1 3 1 1 1 1 1 1 1 1 1 1 1 1 2 1 1 2
1 2 1 1 2 1 3 2 1 1 1 1 1 1 2 1 1 1 1
[325] 3 3 1 1 1 1 1 2 1 2 1 3 1 1 1 1 1 1 1 2 1 1 1 3 1 2 1 2 1 2 1 1 1 1 1
1 1 1 1 1 1 1 1 1 1 3 1 1 1 1 1 1 3
[379] 1 1 3 1 3 1 2 1 1 1 1 1 1 1 1 3 1 1 1 1 1 1 1 3 3 3 1 1 3 2 1 1 1 1 1
1 1 1 1 2 1 1 1 3 1 1 1 1 3 1 1 1 1
[433] 1 1 1 3 3 2 1 1
```

From the output k-means, we can observe and infer several things about the output clusters. Obviously, three clusters are formed and this is in line with our k parameter that was passed to the algorithm. We see that the first cluster has 330 observations in it, and the second and third clusters are small with just 50 and 60 observations. The k-means output also provides us with the cluster centroids. The **centroid** is a representative of all points in a particular cluster. As it is not feasible to study each of the individual observations assigned to a cluster and determine the business characteristics of the cluster, the cluster centroid may be used as a pseudo for the points in the cluster. The cluster centroid helps us to quickly arrive at a conclusion in terms of the definition of contents of the cluster. The k-means output also produced the cluster assignment for each observation. Each of the observations in our wholesale dataset is assigned to one of the three clusters (1,2,3).

It is possible to view the clustering results by using the `fviz_cluster()` function available in the `factoextra` library. The function provides a nice illustration of the clusters. If there are more than two dimensions (variables), `fviz_cluster` will perform **principal component analysis** (**PCA**) and plot the observations based on the first two principal components that explain the majority of the variance. The clusters visualization can be created though the following code:

```
library(factoextra)
fviz_cluster(kmout,data=cust_data)
```

This will give the following graph as output:

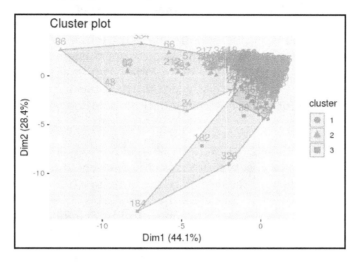

Working mechanics of the k-means algorithm

The execution of the k-means algorithm involves the following steps:

1. Randomly select k observations from the dataset as the initial cluster centroids.
2. For each observation in the dataset, perform the following:
    1. Compute the distance between the observation and each of the cluster centroids.
    2. Identify the cluster centroid that has minimum distance with the observation.
    3. Assign the observation to such closest centroid.

3. With all points assigned to one of the cluster centroids, compute new cluster centroids. This can be done by taking the mean of all the points assigned to a cluster.
4. Perform *step 2* and *step 3* repeatedly until the cluster centroids (mean) do not change or until a user-defined number of iterations is reached.

One key thing to note in k-means is that the cluster centroids in the initial step are selected randomly and the initial cluster assignments are done based on the distance between the actual observations and the randomly-picked cluster centroids. This essentially means that if we were to pick observations as cluster centroids in the initial step other than the observations that were chosen, we would obtain different clusters than the one we have obtained. In technical terms, this is called a **non-globally-optimal solution** or a **locally-optimal solution**. The `cluster` library's k-means function has the `nstart` option, which works around this problem of the non-globally-optimal solution encountered with the k-means algorithm.

The `nstart` option enables the algorithm to try several random starts (instead of just one) by drawing a number of center observations from the datasets. It then checks the cluster sum of squares and proceeds with the best start, resulting in a more stable output. In our case, we set the `nstart` value as `50`, therefore the best start is chosen by k-means post checking it with 50 random initial sets of cluster centroids. The following diagram depicts the high-level steps involved in the k-means clustering algorithm:

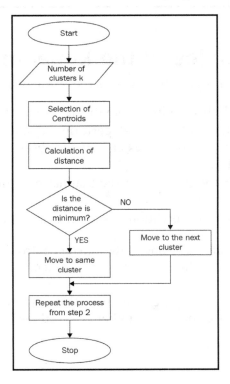

Steps in k-means clustering

In supervised ML methods, such as classification, we have ground truth, therefore, we will be able to able to compare our predictions with the ground truth and measure to report the performance of our classification. Unlike the supervised ML method, in clustering, we do not have any ground truth. Therefore, computing the performance measurement with respect to clustering is a challenge.

As an alternative to performance measurement, we use a pseudo-measure called **cluster quality**. The cluster quality is generally computed through measures known as intra-cluster distance and inter-cluster distance, which are illustrated in the following diagram:

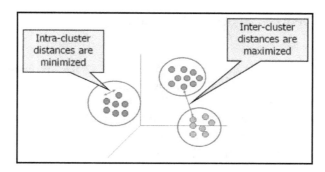

Intra-cluster distance and inter-cluster distance defined

The goal of the clustering task is to obtain good-quality clusters. Clusters are termed as **high-quality clusters** if the distance within the observations is minimum and the distance between the clusters themselves is maximum.

There are multiple ways inter-cluster and intra-cluster distances can be measured:

- **Intra-cluster**: This distance can be measured as (sum, minimum, maximum, or mean) of the (absolute/squared) distance between all pairs of points in the cluster (or) *diameter*–two farthest points (or) between the centroid and all points in the cluster.
- **Inter-cluster**: This distance is measured as sum of the (squared) distance between all pairs of clusters, where distance between two clusters itself is computed as one of the following:
  - Distance between their centroids
  - Distance between farthest pair of points
  - Distance between the closest pair of points belonging to the clusters

Unfortunately, there is no way to pinpoint the preferred inter-cluster distance and intra-cluster distance values. The **Silhouette index** is one metric that is based on inter-cluster distance and intra-cluster distance that can be readily computed and easily interpreted.

The Silhouette index is computed using the mean intra-cluster distance, a, and the mean nearest-cluster distance, b, for each of the observations participating in the clustering exercise. The Silhouette index for an observation is given by the following formula:

$$(b - a)/max(a, b)$$

Here, b is the distance between an observation and the nearest cluster that the observation is not a part of.

Silhouette index value ranges between [-1, 1]. A value of +1 for an observation indicates that the observation is far away from its neighboring cluster and it is very close to the cluster it is assigned to. Similarly, a value of -1 tells us that the observation is close to its neighboring cluster than to the cluster it is assigned to. A value of 0 means it's at the boundary of the distance between the two clusters. A value of +1 is ideal and -1 is the least preferred. Hence, the higher the value, the better the quality of the cluster.

The `cluster` library offers the Silhouette function, which can be readily used on our k-means clustering output to understand the quality of the clusters that were formed. The following code computes the Silhouette index for our three clusters:

```
# computing the silhouette index for the clusters
si <- silhouette(kmout$cluster, dist(cust_data, "euclidean"))
# printing the summary of the computed silhouette index
print(summary(si))
```

This will give us the following output:

```
Silhouette of 440 units in 3 clusters from silhouette.default(x =
kmout$cluster, dist = dist(cust_data, from "euclidean")) :
 Cluster sizes and average silhouette widths:
       60 50 330
0.2524346 0.1800059 0.5646307
Individual silhouette widths:
   Min. 1st Qu. Median Mean 3rd Qu. Max.
-0.1544 0.3338 0.5320 0.4784 0.6743 0.7329
```

As we have seen, the Silhouette index can range from -1 to +1, and the latter is preferred. From the output, we see that the clusters are all good quality clusters, as the average width is a positive number closer to 1 than -1.

In fact, the Silhouette index can be used not just to measure the quality of clusters formed but also to compute the k-value. Similar to Elbow method, we can iterate through multiple values of k and then identify the k that yields the maximum Silhouette index values across the clusters. Clustering can then be performed using the k that was identified.

There are numerous cluster-quality measures described in the literature. The Silhouette index is just one measure we covered in this chapter because of its popularity in the ML community. The `clusterCrit` library offers a wide range of indices to measure the quality of clusters. We are not going to explore the other cluster-quality metrics here, but interested readers should refer to this library for further information on how to compute cluster quality.

So far, we have covered the k-means clustering algorithm to identify the clusters, but the original segmentation task we started with does not end here. Segmentation further spans to the task of understanding what each of the clusters formed from the clustering exercise mean to businesses. For example, we take our cluster centroids obtained from k-means and an attempt is made to identify what these are:

```
Fresh Milk Grocery Frozen Detergents_Paper Delicatessen
1 8253.47 3824.603 5280.455 2572.661 1773.058 1137.497
2 8000.04 18511.420 27573.900 1996.680 12407.360 2252.020
3 35941.40 6044.450 6288.617 6713.967 1039.667 3049.467
```

Here are some sample insights into each cluster:

- Cluster 1 is low spenders (average spending: 22,841.744), with the majority of spending allocated to the fresh category
- Cluster 2 is high spenders (average spending: 70,741.42), with the majority of spending in the grocery category
- Cluster 3 is medium spenders (average spending : 59,077.568), with the majority of spending in the fresh category

Now, based on the business objective, one or more clusters may be selected to target. For example, if the objective is to have high spenders spend more, promotions may be rolled out to cluster 2 individuals with spending in the `Frozen` and `Delicatessen` products less than the centroid values (that is, `Frozen: 1,996.680` and `Delicatessen: 2,252.020`).

Identifying the customer segments in the wholesale customer data using DIANA

Hierarchical clustering algorithms are a good choice when we don't necessarily have circular (or hyperspherical) clusters in the data, and we essentially don't know the number of clusters in advance. With hierarchical clustering algorithm, unlike the flat or partitioning algorithms, there is no requirement to decide and pass the number of clusters to be formed prior to applying the algorithm on the dataset.

Hierarchical clustering results in a dendogram (tree diagram) that can be visually verified to easily determine the number of clusters. Visual verification enables us to perform cuts in the dendrogram at suitable places.

The results produced by this type of clustering algorithm are reproducible as the algorithm is not sensitive to the choice of the distance metric. In other words, irrespective of the distance metric chosen, we will get the same results. This type of clustering is also suitable for datasets of a higher complexity (quadratic) and in particular for exploring the hierarchical relationships that exist between the clusters.

Divisive hierarchical clustering, also known as **DIvisive ANAlysis (DIANA)**, is a hierarchical clustering algorithm that follows a top-down approach to identify clusters in a given dataset. Here are the steps in DIANA to identify the clusters:

1. All observations of the dataset are assigned to the root, so in the initial step only a single cluster is formed.
2. In each iteration, the most heterogeneous cluster is partitioned into two.
3. **Step 2** is repeated until all the observations are in their own cluster:

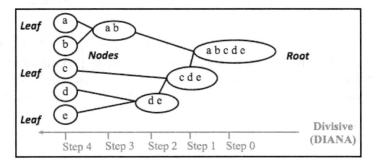

Working of divisive hierarchical clustering algorithm

One obvious question that comes up is about the technique used by the algorithm to split the cluster into two. The answer is that it is performed according to some (dis)similarity measure. The Euclidean distance is used to measure the distance between two given points. This algorithm works by splitting the data on the basis of the farthest-distance measure of all the pairwise distances between the data points. Linkage defines the specific details of fartherness of the data points. The next figure illustrates the various linkages considered by DIANA for splitting the clusters. Here are some of the distances considered to split the groups:

- **Single-link**: Nearest distance or single linkage
- **Complete-link**: Farthest distance or complete linkage
- **Average-link**: Average distance or average linkage
- **Centroid-link**: Centroid distance
- **Ward's method**: Sum of squared `euclidean` distance is minimized

Take a look at the following diagram to better understand the preceding distances:

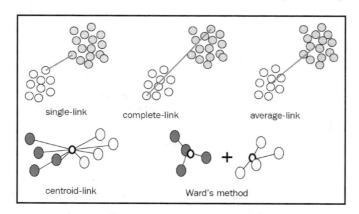

Illustration depicting various linkage types used by DIANA

Generally, the linkage type to be used is passed as a parameter to the clustering algorithm. The `cluster` library offers the `diana` function to perform clustering. Let's apply it on our wholesale dataset with the following code:

```
# setting the working directory to a folder where dataset is located
setwd('/home/sunil/Desktop/chapter5/')
# reading the dataset to cust_data dataframe
cust_data = read.csv(file='Wholesale_customers_ data.csv', header = TRUE)
# removing the non-required columns
cust_data<-cust_data[,c(-1,-2)]
# including the cluster library so as to make use of diana function
```

```
library(cluster)
# Compute diana()
cust_data_diana<-diana(cust_data, metric = "euclidean",stand = FALSE)
# plotting the dendogram from diana output
pltree(cust_data_diana, cex = 0.6, hang = -1,
       main = "Dendrogram of diana")
# Divise coefficient; amount of clustering structure found
print(cust_data_diana$dc)
```

This will give us the following output:

```
> print(cust_data_diana$dc)
[1] 0.9633628
```

Take a look at the following output:

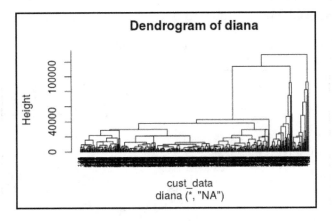

The `plot.hclust()` and `plot.dendrogram()` functions may also be used on the DIANA clustering output. `plot.dendrogram()` yields the dendogram that follows the natural structure of the splits as done by the DIANA algorithm. Use the following code to generate the dendrogram:

```
plot(as.dendrogram(cust_data_diana), cex = 0.6,horiz = TRUE)
```

This will give the following output:

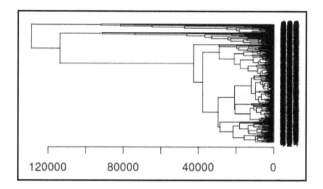

In the dendrogram output, each leaf that appears on the right relates to each observation in the dataset. As we traverse from right to left, observations that are similar to each other are grouped into one branch, which are themselves fused at a higher level.

The higher level of the fusion, provided on the horizontal axis, indicates the similarity between two observations. The higher the fusion, the more similar the observations are. It may be noted that conclusions about the proximity of two observations can be drawn only based on the level where branches containing those two observations are first fused. In order to identify clusters, we can cut the dendrogram at a certain level. The level at which the cut is made defines the number of clusters obtained.

We can make use of the `cutree()` function to obtain the cluster assignment for each of the observations in our dataset. Execute the following code to obtain the clusters and review the clustering output:

```
# obtain the clusters through cuttree
# Cut tree into 3 groups
grp <- cutree(cust_data_diana, k = 3)
# Number of members in each cluster
table(grp)
# Get the observations of cluster 1
rownames(cust_data)[grp == 1]
```

This will give the following output:

```
> table(grp)
grp
  1 2 3
364 44 32
> rownames(cust_data)[grp == 1]
   [1] "1"  "2"  "3"  "4"  "5"  "6"  "7"  "8"  "9"  "11"  "12"  "13"  "14"  "15"  "16"
```

```
"17" "18" "19"
  [19] "20" "21" "22" "25" "26" "27" "28" "31" "32" "33" "34" "35" "36" "37"
"38" "41" "42" "43"
  [37] "45" "49" "51" "52" "54" "55" "56" "58" "59" "60" "61" "63" "64" "65"
"67" "68" "69" "70"
  [55] "71" "72" "73" "74" "75" "76" "77" "79" "80" "81" "82" "83" "84" "85"
"89" "90" "91" "92"
  [73] "94" "95" "96" "97" "98" "99" "100" "101" "102" "103" "105" "106"
"107" "108" "109" "111" "112" "113"
  [91] "114" "115" "116" "117" "118" "119" "120" "121" "122" "123" "124"
"127" "128" "129" "131" "132" "133" "134"
 [109] "135" "136" "137" "138" "139" "140" "141" "142" "144" "145" "147"
"148" "149" "151" "152" "153" "154" "155"
 [127] "157" "158" "159" "160" "161" "162" "163" "165" "167" "168" "169"
"170" "171" "173" "175" "176" "178" "179"
 [145] "180" "181" "183" "185" "186" "187" "188" "189" "190" "191" "192"
"193" "194" "195" "196" "198" "199" "200"
 [163] "203" "204" "205" "207" "208" "209" "211" "213" "214" "215" "216"
"218" "219" "220" "221" "222" "223" "224"
 [181] "225" "226" "227" "228" "229" "230" "231" "232" "233" "234" "235"
"236" "237" "238" "239" "241" "242" "243"
 [199] "244" "245" "246" "247" "248" "249" "250" "251" "253" "254" "255"
"257" "258" "261" "262" "263" "264" "265"
 [217] "266" "268" "269" "270" "271" "272" "273" "275" "276" "277" "278"
"279" "280" "281" "282" "284" "287" "288"
 [235] "289" "291" "292" "293" "294" "295" "296" "297" "298" "299" "300"
"301" "303" "304" "306" "308" "309" "311"
 [253] "312" "314" "315" "316" "317" "318" "319" "321" "322" "323" "324"
"325" "327" "328" "329" "330" "331" "333"
 [271] "335" "336" "337" "338" "339" "340" "341" "342" "343" "345" "346"
"347" "348" "349" "351" "353" "355" "356"
 [289] "357" "358" "359" "360" "361" "362" "363" "364" "365" "366" "367"
"368" "369" "370" "372" "373" "374" "375"
 [307] "376" "377" "379" "380" "381" "382" "384" "385" "386" "387" "388"
"389" "390" "391" "392" "393" "394" "395"
 [325] "396" "397" "398" "399" "400" "401" "402" "403" "404" "405" "406"
"407" "409" "410" "411" "412" "413" "414"
 [343] "415" "416" "417" "418" "420" "421" "422" "423" "424" "425" "426"
"427" "429" "430" "431" "432" "433" "434"
 [361] "435" "436" "439" "440"
```

We can also visualize the clustering output through the `fviz_cluster` function in the `factoextra` library. Use the following code to get the required visualization:

```
library(factoextra)
fviz_cluster(list(data = cust_data, cluster = grp))
```

This will give you the following output:

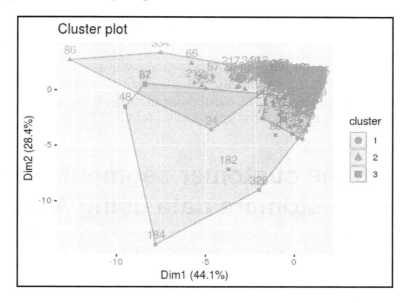

It is also possible to color-code the clusters within the dendogram itself. This can be accomplished with the following code:

```
plot(as.hclust(cust_data_diana))
rect.hclust(cust_data_diana, k = 4, border = 2:5)
```

This will give the following output:

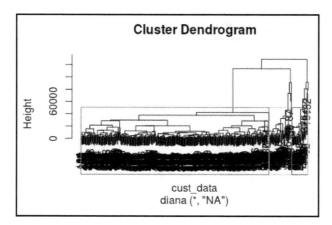

Now that the clusters are identified, the steps we discussed to evaluate the cluster quality (through the Silhouette index) apply here as well. As we have already covered this topic under the k-means clustering algorithm, we are not going to repeat the steps here. The code and interpretation of the output remains the same as what was discussed under k-means.

As discussed earlier, the cluster's output is not the final point to customer segmentation exercise we have on hand. Similar to the discussion we had on under the k-means algorithm, we could analyze the DIANA cluster output to identify meaningful segments so as to roll out business objectives to those specifically-identified segments.

Identifying the customer segments in the wholesale customers data using AGNES

AGNES is the reverse of DIANA in the sense that it follows a bottom-up approach to clustering the dataset. The following diagram illustrates the working principle of the AGNES algorithm for clustering:

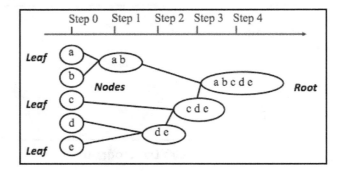

Working of agglomerative hierarchical clustering algorithm

Except for the bottom-up approach followed by AGNES, the implementation details behind the algorithm are the same as for DIANA; therefore, we won't repeat the discussion of the concepts here. The following code block clusters our wholesale dataset into three clusters with AGNES; it also creates a visualization of the clusters thus formed:

```
# setting the working directory to a folder where dataset is located
setwd('/home/sunil/Desktop/chapter5/')
# reading the dataset to cust_data dataframe
cust_data = read.csv(file='Wholesale_customers_ data.csv', header = TRUE)
# removing the non-required columns
cust_data<-cust_data[,c(-1,-2)]
# including the cluster library so as to make use of agnes function
library(cluster)
# Compute agnes()
cust_data_agnes<-agnes(cust_data, metric = "euclidean",stand = FALSE)
# plotting the dendogram from agnes output
pltree(cust_data_agnes, cex = 0.6, hang = -1,
       main = "Dendrogram of agnes")
# agglomerative coefficient; amount of clustering structure found
print(cust_data_agnes$ac)
plot(as.dendrogram(cust_data_agnes), cex = 0.6,horiz = TRUE)
# obtain the clusters through cuttree
# Cut tree into 3 groups
grp <- cutree(cust_data_agnes, k = 3)
# Number of members in each cluster
table(grp)
# Get the observations of cluster 1
rownames(cust_data)[grp == 1]
# visualization of clusters
library(factoextra)
fviz_cluster(list(data = cust_data, cluster = grp))
library(factoextra)
fviz_cluster(list(data = cust_data, cluster = grp))
plot(as.hclust(cust_data_agnes))
rect.hclust(cust_data_agnes, k = 3, border = 2:5)
```

This is the output that you will obtain:

```
[1] 0.9602911
> plot(as.dendrogram(cust_data_agnes), cex = 0.6,horiz = FALSE)
```

Take a look at the following screenshot:

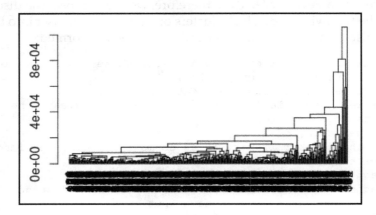

Take a look at the following code block:

```
> grp <- cutree(cust_data_agnes, k = 3)
> # Number of members in each cluster
> table(grp)
grp
  1 2 3
434 5 1
> rownames(cust_data)[grp == 1]
  [1] "1"   "2"   "3"   "4"   "5"   "6"   "7"   "8"   "9"   "10"  "11"  "12"  "13"  "14"  "15"
"16"  "17"  "18"
 [19] "19"  "20"  "21"  "22"  "23"  "24"  "25"  "26"  "27"  "28"  "29"  "30"  "31"  "32"
"33"  "34"  "35"  "36"
 [37] "37"  "38"  "39"  "40"  "41"  "42"  "43"  "44"  "45"  "46"  "47"  "49"  "50"  "51"
"52"  "53"  "54"  "55"
 [55] "56"  "57"  "58"  "59"  "60"  "61"  "63"  "64"  "65"  "66"  "67"  "68"  "69"  "70"
"71"  "72"  "73"  "74"
 [73] "75"  "76"  "77"  "78"  "79"  "80"  "81"  "82"  "83"  "84"  "85"  "88"  "89"  "90"
"91"  "92"  "93"  "94"
 [91] "95"  "96"  "97"  "98"  "99"  "100" "101" "102" "103" "104" "105" "106"
"107" "108" "109" "110" "111" "112"
[109] "113" "114" "115" "116" "117" "118" "119" "120" "121" "122" "123"
"124" "125" "126" "127" "128" "129" "130"
[127] "131" "132" "133" "134" "135" "136" "137" "138" "139" "140" "141"
"142" "143" "144" "145" "146" "147" "148"
[145] "149" "150" "151" "152" "153" "154" "155" "156" "157" "158" "159"
"160" "161" "162" "163" "164" "165" "166"
[163] "167" "168" "169" "170" "171" "172" "173" "174" "175" "176" "177"
"178" "179" "180" "181" "183" "184" "185"
[181] "186" "187" "188" "189" "190" "191" "192" "193" "194" "195" "196"
"197" "198" "199" "200" "201" "202" "203"
```

```
[199] "204" "205" "206" "207" "208" "209" "210" "211" "212" "213" "214"
"215" "216" "217" "218" "219" "220" "221"
[217] "222" "223" "224" "225" "226" "227" "228" "229" "230" "231" "232"
"233" "234" "235" "236" "237" "238" "239"
[235] "240" "241" "242" "243" "244" "245" "246" "247" "248" "249" "250"
"251" "252" "253" "254" "255" "256" "257"
[253] "258" "259" "260" "261" "262" "263" "264" "265" "266" "267" "268"
"269" "270" "271" "272" "273" "274" "275"
[271] "276" "277" "278" "279" "280" "281" "282" "283" "284" "285" "286"
"287" "288" "289" "290" "291" "292" "293"
[289] "294" "295" "296" "297" "298" "299" "300" "301" "302" "303" "304"
"305" "306" "307" "308" "309" "310" "311"
[307] "312" "313" "314" "315" "316" "317" "318" "319" "320" "321" "322"
"323" "324" "325" "326" "327" "328" "329"
[325] "330" "331" "332" "333" "335" "336" "337" "338" "339" "340" "341"
"342" "343" "344" "345" "346" "347" "348"
[343] "349" "350" "351" "352" "353" "354" "355" "356" "357" "358" "359"
"360" "361" "362" "363" "364" "365" "366"
[361] "367" "368" "369" "370" "371" "372" "373" "374" "375" "376" "377"
"378" "379" "380" "381" "382" "383" "384"
[379] "385" "386" "387" "388" "389" "390" "391" "392" "393" "394" "395"
"396" "397" "398" "399" "400" "401" "402"
[397] "403" "404" "405" "406" "407" "408" "409" "410" "411" "412" "413"
"414" "415" "416" "417" "418" "419" "420"
[415] "421" "422" "423" "424" "425" "426" "427" "428" "429" "430" "431"
"432" "433" "434" "435" "436" "437" "438"
[433] "439" "440"
```

Execute the following command:

```
> fviz_cluster(list(data = cust_data, cluster = grp))
```

The preceding command generates the following output:

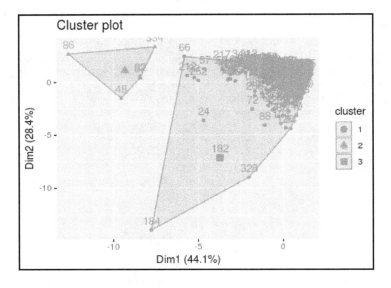

Take a look at the following commands:

```
> plot(as.hclust(cust_data_agnes))
> rect.hclust(cust_data_agnes, k = 3, border = 2:5)
```

The preceding commands generate the following output:

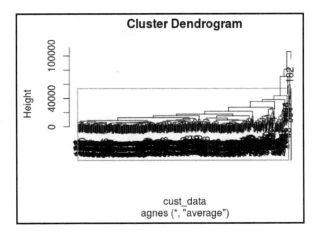

We can see from the AGNES clustering output that a large number of observations from the dataset are assigned to one cluster and very few observations were assigned to the other clusters. This is not a great output for our segmentation downstream exercise. To obtain better cluster assignments, you could try using other cluster-linkage methods aside from the default average linkage method currently used by the AGNES algorithm.

Summary

In this chapter, we learned about the concept of segmentation and its association with clustering, an ML unsupervised learning technique. We made use of the wholesale dataset available from the UCI repository and implemented clustering using the k-means, DIANA, and AGNES algorithms. During the course of this chapter, we also studied various aspects related to clustering, such as tendency to cluster, distance, linkage measures, and methods to identify the right number of clusters, and measuring the output of clustering. We also explored making use of the clustering output for customer-segmentation purposes.

Can computers see and identify objects and living creatures like humans do? Let's explore the answer to this question in the next chapter.

6
Image Recognition Using Deep Neural Networks

In 1966, Professor Seymour Papert at MIT conceptualized an ambitious summer project titled *The Summer Vision Project*. The task for the graduate student was to *plug a camera into a computer and enable it to understand what it sees*! I am sure it would have been super-difficult for the graduate student to have finished this project, as even today the task remains half complete.

A human being, when they look outside, is able to recognize the objects that they see. Without thinking, they are able to classify a cat as a cat, a dog as a dog, a plant as a plant, an animal as an animal—this is happening because the human brain draws knowledge from its extensive prelearned database. After all, as human beings, we have millions of years' worth of evolutionary context that enables us draw inferences from the thing that we see. Computer vision deals with replicating the human vision processes so as to pass them on to machines and automate them.

This chapter is all about learning the theory and implementation of computer vision through **machine learning** (**ML**). We will build a feedforward deep learning network and **LeNet** to enable handwritten digit recognition. We will also build a project that uses a pretrained Inception-BatchNorm network to identify objects in an image. We will cover the following topics as we progress in this chapter:

- Understanding computer vision
- Achieving computer vision with deep learning
- Introduction to the MNIST dataset
- Implementing a deep learning network for handwritten digit recognition
- Implementing computer vision with pretrained models

Technical requirements

For the projects covered in this chapter, we'll make use of a very popular open dataset called MNIST. We'll use **Apache MXNet**, a modern open source deep learning software framework to train and deploy the required deep neural networks.

Understanding computer vision

In today's world, we have advanced cameras that are very successful at mimicking how a human eye captures light and color; but image-capturing in the right way is just stage one in the whole image-comprehension aspect. Post image-capturing, we will need to enable technology that interprets what has been captured and build context around it. This is what the human brain does when the eyes see something. Here comes the huge challenge: we all know that computers see images as huge piles of integer values that represent intensities across a spectrum of colors, and of course, computer have no context associated with the image itself. This is where ML comes into play. ML allows us to train a context for a dataset such that it enables computers to understand what objects certain sequences of numbers actually represent.

Computer vision is one of the emerging areas where ML is applied. It can be used for several purposes in various domains, including healthcare, agriculture, insurance, and the automotive industry. The following are some of its most popular applications:

- Detecting diseases from medical images, such as CT scan/MRI scan images
- Identifying crop diseases and soil quality to support a better crop yield
- Identifying oil reserves from satellite images
- Self-driving cars
- Monitoring and managing skin condition for psoriasis patients
- Classifying and distinguishing weeds from crops
- Facial recognition
- Extracting information from personal documents, such as passports and ID cards
- Detecting terrain for drones and airplanes
- Biometrics
- Public surveillance
- Organizing personal photos
- Answreing visual questions

This is just the tip of the iceberg. It's not an overstatement to say that there is no domain where we cannot find an application for computer vision. Therefore, computer vision is a key area for ML practitioners to focus on.

Achieving computer vision with deep learning

To start with, let's understand the term **deep learning**. It simply means **multilayered neural networks**. The multiple layers enable deep learning to be an enhanced and powerful form of a neural network. **Artificial neural networks** (**ANNs**) have been in existence since the 1950s. They have always been designed with two layers; however, deep learning models are built with multiple hidden layers. The following diagram shows a hypothetical deep learning model:

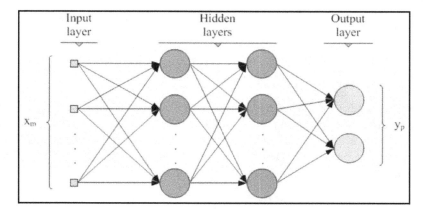

Deep learning model—High level architecture

Neural networks are heavy on computation, therefore the **central processing unit** (**CPU**) that can be enabled with a maximum of 22 cores is generally thought of as an infrastructure blocker until recently. This infrastructure limitation also limited the usage of neural networks to solve real-world problems. However, recently, the availability of a **graphical processing unit** (**GPU**) with thousands of cores enabled has exponentially powerful computation possibilities when compared to CPUs. This gave a huge push to the usage of deep learning models.

Data comes in many forms, such as tables, sounds, HTML files, TXT files, and images. Linear models do not generally learn from non-linear data. Non-linear algorithms, such as decision trees and gradient-boosting machines, also do not learn well from this kind of data. One the other hand, deep learning models that create non-linear interactions among the features give better solutions with non-linear data, so they have become the preferred models in the ML community.

A deep learning model consists of a chain of interconnected neurons that creates the neural architecture. Any deep learning model will have an input layer, two or more hidden layers (middle layers), and an output layer. The input layer consists of neurons equal to the number of input variables in the data. Users can decide on the number of neurons and the number of hidden layers that a deep learning network should have. Generally, it is something that is optimized by the user building the network through a cross-validation strategy. The choice of the number of neurons and the number of hidden layers represents the challenge of the researcher. The number of neurons in the output layer is decided based on the outcome of the problem. For example, one output neuron in case it is regression, for a classification problem the output neurons is equal to the number of classes involved in the problem on-hand.

Convolutional Neural Networks

There are multiple types of deep learning algorithms, the one we generally use in computer vision is called a **Convolutional Neural Network** (**CNN**). CNNs break down images into small groups of pixels and then run calculations on them by applying filters. The result is then compared against pixel matrices they already know about. This helps CNNs to come up with a probability for the image belonging to one of the known classes.

In the first few layers, the CNN identifies shapes, such as curves and rough edges, but after several convolutions, they are able to recognize objects such as animals, cars, and humans.

When the CNN is first built for the available data, the filter values of the network are randomly initialized and so the predictions it produce are mostly false. But then it keeps comparing its own predictions on labeled datasets to the actual ones, updating the filter values and improving performance of the CNN with each iteration.

Layers of CNNs

A CNN consists of an input and an output layer; it also has various hidden layers. The following are the various hidden layers in a CNN:

- **Convolution**: Assume that we have an image represented as pixels, a convolution is something where we have a little matrix nearly always 3 x 3 in deep learning and multiply every element of the matrix by every element of 3 x 3 section of the image and then add them all together to get the result of that convolution at one point. The following diagram illustrates the process of convolution on a pixel:

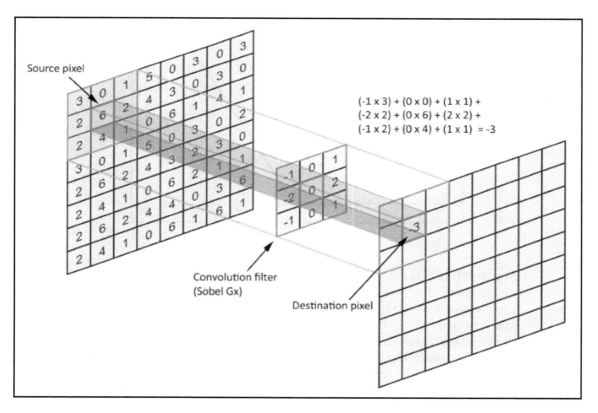

Convolution application on an image

- **Rectified Linear Unit (ReLU)**: A non-linear activation that throws away the negatives in an input matrix. For example, let's assume we have a 3 x 3 matrix with negative numbers, zeros, and positive numbers as values in the cells of the matrix. Given this matrix as input to ReLU, it transforms all negative numbers in the matrix to zeros and returns the 3 x 3 matrix. ReLU is an activation function that can be defined as part of the CNN architecture. The following diagram demonstrates the function of ReLU in CNNs:

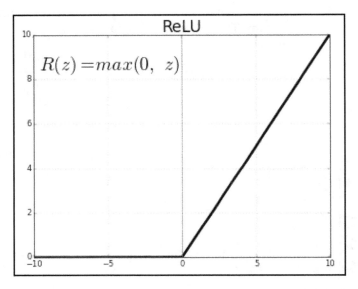

Rectified Linear Unit (ReLU) in CNNs

- **Max pooling**: Max pooling is something that can be set as a layer in the CNN architecture. It allows to identify if the specific characteristic is present in the previous level. It replaces the highest value in an input matrix with the maximum and gives the output. Let's consider an example, with a 2 x 2 max pooling layer, given a 4 x 4 matrix as input, the max pooling layer replaces each 2 x 2 in the input matrix with the highest value among the four cells. The output matrix thus obtained is non-overlapping and it's an image representation with a reduced resolution. The following diagram illustrates the functionality of max pooling in a CNN:

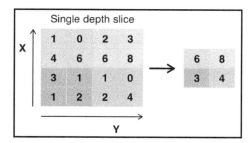

Functionality of max pooling layer in CNNs

There are various reasons to apply max pooling, such as to reduce the amount of parameters and computation load, to eliminate overfitting, and, most importantly, to force the neural network to see the larger picture, as in previous layers it was focused on seeing bits and pieces of the image.

- **Fully-connected layer**: Also known as a **dense layer**, this involves a linear operation on the layer's input vector. The layer ensures every input is connected to every output by a weight.
- **Softmax**: An activation function that is generally applied at the last layer of the deep neural network. In a multiclass classification problem, we require the fully-connected output of a deep learning network to be interpreted as a probability. The total probability of a particular observation in data (for all classes) should add up to 1, and the probability of the observation belonging to each class should range between 0 and 1. Therefore, we transform each output of the fully-connected layer as a portion of a total sum. However, instead of simply doing the standard proportion, we apply this non-linear exponential function for a very specific reason: we would like to make our highest output as close to 1 as possible and our lower output as close to 0. Softmax does this by pushing the true linear proportions closer to either 1 or 0.

The following diagram illustrates the softmax activation function:

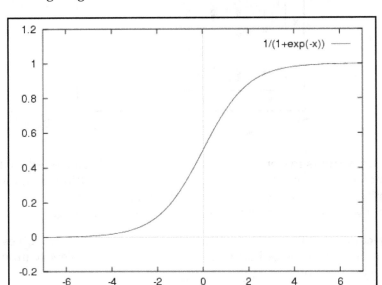

Softmax activation function

- **Sigmoid**: This is similar to softmax, except that it is applied to a binary classification, such as cats versus dogs. With this activation function, the class to which the observation belongs is assigned a higher probability compared to the other class. Unlike softmax, the probabilities do not have to add up to 1.

Introduction to the MXNet framework

MXNet is a super-powerful open source deep learning framework that is built to ease the development of deep learning algorithms. It is used to define, train, and deploy deep neural networks. MXNet is lean, flexible, and ultra-scalable, that is, it allows fast model training and supports a flexible programming model with multiple languages. The problem with existing deep learning frameworks, such as Torch7, Theano, and Caffe, is that users need to learn another system or a different programming flavor.

However, MXNet resolves this issue by supporting multiple languages, such as C++, Python, R, Julia, and Perl. This eliminates the need for users to learn a new language; therefore, they can use the framework and simplify network definitions. MXNet models are able to fit in small amounts of memory and they can be trained on CPUs, GPUs, and on multiple machines with ease. The `mxnet` package is readily available for the R language and the details of the install can be looked up in **Apache Incubator** at `https://mxnet.incubator.apache.org`.

Understanding the MNIST dataset

Modified National Institute of Standards and Technology (MNIST) is a dataset that contains images of handwritten digits. This dataset is pretty popular in the ML community for implementing and testing computer vision algorithms. The MNIST dataset is an open dataset made available by Professor Yann LeCun at `http://yann.lecun.com/exdb/mnist/`, where separate files that represent the training dataset and test dataset are available. The labels corresponding to the test and training datasets are also available as separate files. The training dataset has 60,000 samples and the test dataset has 10,000 samples.

The following diagram shows some sample images from the MNIST dataset. Each of the images also comes with a label indicating the digit shown in the following screenshot:

Sample images from MNIST dataset

The labels for the images shown in the preceding diagram are **5**, **0**, **4**, and **1**. Each image in the dataset is a grayscale image and is represented in 28 x 28 pixels. A sample image represented with pixels is shown in the following screenshot:

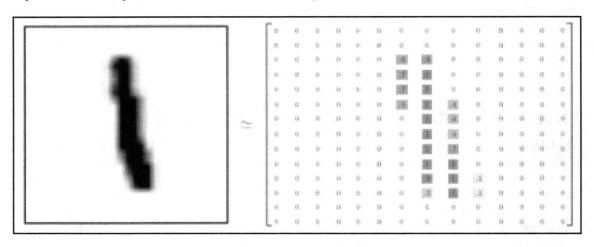

Sample image from MNIST dataset represented with 28 * 28 pixels

It is possible to flatten the 28 x 28 pixel matrix and represent it as a vector of 784 pixel values. Essentially, the training dataset is a 60,000 x 784 matrix that could be used with ML algorithms. The test dataset is a 10,000 x 784 matrix. The training and test datasets may be downloaded from the source with the following code:

```
# setting the working directory where the files need to be downloaded
setwd('/home/sunil/Desktop/book/chapter 6/MNIST')
# download the training and testing dataset from source
download.file("http://yann.lecun.com/exdb/mnist/train-images-idx3-ubyte.gz"
,"train-images-idx3-ubyte.gz")
download.file("http://yann.lecun.com/exdb/mnist/train-labels-idx1-ubyte.gz"
,"train-labels-idx1-ubyte.gz")
download.file("http://yann.lecun.com/exdb/mnist/t10k-images-idx3-ubyte.gz",
"t10k-images-idx3-ubyte.gz")
download.file("http://yann.lecun.com/exdb/mnist/t10k-labels-idx1-ubyte.gz",
"t10k-labels-idx1-ubyte.gz")
# unzip the training and test zip files that are downloaded
R.utils::gunzip("train-images-idx3-ubyte.gz")
R.utils::gunzip("train-labels-idx1-ubyte.gz")
R.utils::gunzip("t10k-images-idx3-ubyte.gz")
R.utils::gunzip("t10k-labels-idx1-ubyte.gz")
```

Once the data is downloaded and unzipped, we will see the files in our working directory. However, these files are in binary format and they cannot be directly loaded through the regular `read.csv` command. The following custom function code helps to read the training and test data from the binary files:

```
# function to load the image files
load_image_file = function(filename) {
  ret = list()
  # opening the binary file in read mode
  f = file(filename, 'rb')
  # reading the binary file into a matrix called x
 readBin(f, 'integer', n = 1, size = 4, endian = 'big')
 n = readBin(f, 'integer', n = 1, size = 4, endian = 'big')
 nrow = readBin(f, 'integer', n = 1, size = 4, endian = 'big')
 ncol = readBin(f, 'integer', n = 1, size = 4, endian = 'big')
 x = readBin(f, 'integer', n = n * nrow * ncol, size = 1, signed = FALSE)
  # closing the file
  close(f)
  # converting the matrix and returning the dataframe
  data.frame(matrix(x, ncol = nrow * ncol, byrow = TRUE))
}
# function to load label files
load_label_file = function(filename) {
  # reading the binary file in read mode
  f = file(filename, 'rb')
  # reading the labels binary file into y vector
  readBin(f, 'integer', n = 1, size = 4, endian = 'big')
  n = readBin(f, 'integer', n = 1, size = 4, endian = 'big')
  y = readBin(f, 'integer', n = n, size = 1, signed = FALSE)
  # closing the file
  close(f)
  # returning the y vector
  y
}
```

The functions may be called with the following code:

```
# load training images data through the load_image_file custom function
train = load_image_file("train-images-idx3-ubyte")
# load  test data through the load_image_file custom function
test  = load_image_file("t10k-images-idx3-ubyte")
# load the train dataset labels
train.y = load_label_file("train-labels-idx1-ubyte")
# load the test dataset labels
test.y  = load_label_file("t10k-labels-idx1-ubyte")
```

In RStudio, when we execute the code, we see `train`, `test`, `train.y`, and `test.y` displayed under the **Environment** tab. This confirms that the datasets are successfully loaded and the respective dataframes are created, as shown in the following screenshot:

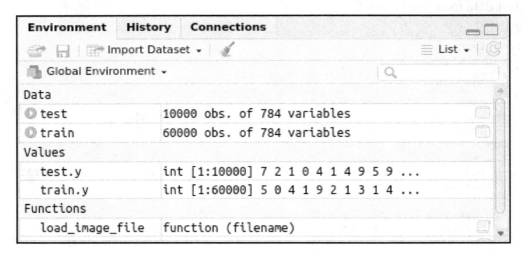

Once the image data is loaded into the dataframe, it is in the form of a series of numbers that represent the pixel values. The following is a helper function that visualizes the pixel data as an image in RStudio:

```
# helper function to visualize image given a record of pixels
show_digit = function(arr784, col = gray(12:1 / 12), ...) {
   image(matrix(as.matrix(arr784), nrow = 28)[, 28:1], col = col, ...)
}
```

The `show_digit()` function may be called like any other R function with the dataframe record number as a parameter. For example, the function in the following code block helps to visualize the 3 record in the training dataset as an image in RStudio:

```
# viewing image corresponding to record 3 in the train dataset
show_digit(train[3, ])
```

This will give the following output:

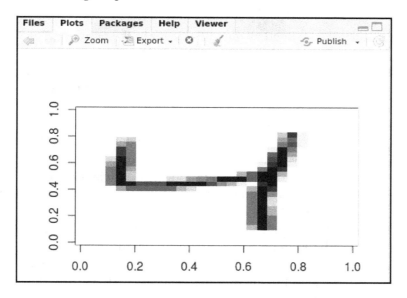

 Dr. David Robinson, in his blog on *Exploring handwritten digit classification: a tidy analysis of the MNIST dataset* (`http://varianceexplained.org/r/digit-eda/`), performed a beautiful exploratory data analysis of the MNIST dataset, which will help you better understand the dataset.

Implementing a deep learning network for handwritten digit recognition

The `mxnet` library offers several functions that enable us to define the layers and activations that comprise the deep learning network. The definition of layers, the usage of activation functions, and the number of neurons to be used in each of the hidden layers is generally termed the **network architecture**. Deciding on the network architecture is more of an art than a science. Often, several iterations of experiments may be needed to decide on the right architecture for the problem. We call it an art as there are no exact rules for finding the ideal architecture. The number of layers, neurons in these layers, and the type of layers are pretty much decided through trial and error.

In this section, we'll build a simple deep learning network with three hidden layers. Here is the general architecture of our network:

1. The input layer is defined as the initial layer in the network. The `mx.symbol.Variable` MXNet function defines the input layer.

2. A fully-connected layer is defined, also called a dense layer, with 128 neurons as the first hidden layer in the network. This can be done using the `mx.symbol.FullyConnected` MXNet function.

3. A ReLU activation function is defined as part of the network. The `mx.symbol.Activation` function helps us to define the ReLU activation function as part of the network.

4. Define the second hidden layer; it is another dense layer with 64 neurons. This can be accomplished through the `mx.symbol.FullyConnected` function, similar to the first hidden layer.

5. Apply a ReLU activation function on the second hidden layer's output. This can be done through the `mx.symbol.Activation` function.

6. The final hidden layer in our network is another fully-connected layer, but with just ten outputs (equal to the number of classes). This can be done through the `mx.symbol.FullyConnected` function as well.

7. The output layer needs to be defined and this should be probabilities of prediction for each class; therefore, we apply softmax at the output layer. The `mx.symbol.SoftmaxOutput` function enables us to configure the softmax in the output.

We are not saying that this is the best network architecture possible for the problem, but this is the network we are going to build to demonstrate the implementation of a deep learning network with MXNet.

Now that we have a blueprint in place, let's delve into coding the network using the following code block:

```
# setting the working directory
setwd('/home/sunil/Desktop/book/chapter 6/MNIST')
# function to load image files
load_image_file = function(filename) {
  ret = list()
  f = file(filename, 'rb')
  readBin(f, 'integer', n = 1, size = 4, endian = 'big')
  n    = readBin(f, 'integer', n = 1, size = 4, endian = 'big')
  nrow = readBin(f, 'integer', n = 1, size = 4, endian = 'big')
  ncol = readBin(f, 'integer', n = 1, size = 4, endian = 'big')
```

```
  x = readBin(f, 'integer', n = n * nrow * ncol, size = 1, signed
= FALSE)
  close(f)
  data.frame(matrix(x, ncol = nrow * ncol, byrow = TRUE))
}
# function to load the label files
load_label_file = function(filename) {
  f = file(filename, 'rb')
  readBin(f, 'integer', n = 1, size = 4, endian = 'big')
  n = readBin(f, 'integer', n = 1, size = 4, endian = 'big')
  y = readBin(f, 'integer', n = n, size = 1, signed = FALSE)
  close(f)
  y }
# loading the image files
train = load_image_file("train-images-idx3-ubyte")
test  = load_image_file("t10k-images-idx3-ubyte")
# loading the labels
train.y = load_label_file("train-labels-idx1-ubyte")
test.y  = load_label_file("t10k-labels-idx1-ubyte")
# lineaerly transforming the grey scale image i.e. between 0 and 255 to # 0
and 1
train.x <- data.matrix(train/255)
test <- data.matrix(test/255)
# verifying the distribution of the digit labels in train dataset
print(table(train.y))
# verifying the distribution of the digit labels in test dataset
print(table(test.y))
```

This will give the following output:

```
train.y
   0    1    2    3    4    5    6    7    8    9
5923 6742 5958 6131 5842 5421 5918 6265 5851 5949

test.y
   0    1    2    3    4    5    6    7    8    9
 980 1135 1032 1010  982  892  958 1028  974 1009
```

Now, define the three layers and start training the network to obtain class probabilities and ensure the results are reproducible using the following code block:

```
# including the required mxnet library
library(mxnet)
# defining the input layer in the network architecture
data <- mx.symbol.Variable("data")
# defining the first hidden layer with 128 neurons and also naming the #
layer as fc1
# passing the input data layer as input to the fc1 layer
```

```
fc1 <- mx.symbol.FullyConnected(data, name="fc1", num_hidden=128)
# defining the ReLU activation function on the fc1 output and also # naming
the layer as ReLU1
act1 <- mx.symbol.Activation(fc1, name="ReLU1", act_type="relu")
# defining the second hidden layer with 64 neurons and also naming the #
layer as fc2
# passing the previous activation layer output as input to the
fc2 layer
fc2 <- mx.symbol.FullyConnected(act1, name="fc2", num_hidden=64)
# defining the ReLU activation function on the fc2 output and also
# naming the layer as ReLU2
act2 <- mx.symbol.Activation(fc2, name="ReLU2", act_type="relu")
# defining the third and final hidden layer in our network with 10
# neurons and also naming the layer as fc3
# passing the previous activation layer output as input to the
fc3 layer
fc3 <- mx.symbol.FullyConnected(act2, name="fc3", num_hidden=10)
# defining the output layer with softmax activation function to obtain #
class probabilities
softmax <- mx.symbol.SoftmaxOutput(fc3, name="sm")
# defining that the experiment should run on cpu
devices <- mx.cpu()
# setting the seed for the experiment so as to ensure that the results #
are reproducible
mx.set.seed(0)
# building the model with the network architecture defined above
model <- mx.model.FeedForward.create(softmax, X=train.x, y=train.y,
ctx=devices, num.round=10, array.batch.size=100,array.layout ="rowmajor",
learning.rate=0.07, momentum=0.9,  eval.metric=mx.metric.accuracy,
initializer=mx.init.uniform(0.07),
epoch.end.callback=mx.callback.log.train.metric(100))
```

This will give the following output:

```
Start training with 1 devices
[1] Train-accuracy=0.885783334343384
[2] Train-accuracy=0.963616671562195
[3] Train-accuracy=0.97510000983874
[4] Train-accuracy=0.980016676982244
[5] Train-accuracy=0.984233343303204
[6] Train-accuracy=0.986883342464765
[7] Train-accuracy=0.98848334223032
[8] Train-accuracy=0.990800007780393
[9] Train-accuracy=0.991300007204215
[10] Train-accuracy=0.991516673564911
```

To make predictions on the test dataset and get the label for each observation in the test dataset, use the following code block:

```
# making predictions on the test dataset
preds <- predict(model, test)
# verifying the predicted output
print(dim(preds))
# getting the label for each observation in test dataset; the
# predicted class is the one with highest probability
pred.label <- max.col(t(preds)) - 1
# observing the distribution of predicted labels in the test dataset
print(table(pred.label))
```

This will give the following output:

```
[1]    10 10000
pred.label
   0    1    2    3    4    5    6    7    8    9
 980 1149 1030 1021 1001  869  960 1001  964 1025
```

Let's check the performance of the model using the following code:

```
# obtaining the performance of the model
print(accuracy(pred.label,test.y))
```

This will give the following output:

```
Accuracy (PCC): 97.73%
Cohen's Kappa: 0.9748
Users accuracy:
   0    1    2    3    4    5    6    7    8    9
98.8 99.6 98.0 97.7 98.3 96.1 97.9 96.3 96.6 97.7
Producers accuracy:
   0    1    2    3    4    5    6    7    8    9
98.8 98.3 98.2 96.7 96.4 98.6 97.7 98.9 97.6 96.2
Confusion matrix
   y
x     0    1    2    3    4    5    6    7    8    9
 0  968    0    1    1    1    2    3    1    2    1
 1    1 1130    3    0    0    1    3    8    1    2
 2    0    1 1011    2    2    0    0   11    3    0
 3    1    2    6  987    0   14    2    2    4    3
 4    1    0    2    1  965    2   10    3    6   11
 5    1    0    0    4    0  857    2    0    3    2
 6    5    2    3    0    4    5  938    0    3    0
 7    0    0    2    2    1    1    0  990    3    2
 8    1    0    4    8    0    5    0    3  941    2
 9    2    0    0    5    9    5    0   10    8  986
```

To visualize the network architecture, use the following code:

```
# Visualizing the network architecture
graph.viz(model$symbol)
```

This will give the following output:

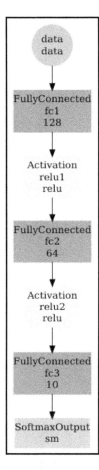

With the simple architecture running for a few minutes on a CPU-based laptop and with minimal effort, we were able to achieve an accuracy of 97.7% on the test dataset. The deep learning network was able to learn to interpret the digits by seeing the images it was given as input. The accuracy of the system can be further improved by altering the architecture or by increasing the number of iterations. It may be noted that, in the earlier experiment, we ran it for 10 iterations.

The number of iterations can simply be amended when model-building through the num.round parameter. There is no hard-and-fast rule in terms of the optimal number of rounds, so this is something to be determined by trial and error. Let's build the model with 50 iterations and observe its impact on performance. The code will remain the same as the earlier project, except with the following amendment to the model-building code:

```
model <- mx.model.FeedForward.create(softmax, X=train.x, y=train.y,
ctx=devices, num.round=50, array.batch.size=100,array.layout ="rowmajor",
learning.rate=0.07, momentum=0.9,  eval.metric=mx.metric.accuracy,
initializer=mx.init.uniform(0.07),
epoch.end.callback=mx.callback.log.train.metric(100))
```

Observe that the num.round parameter is now set to 50, instead of the earlier value of 10.

This will give the following output:

```
[35] Train-accuracy=0.999933333396912
[36] Train-accuracy=1
[37] Train-accuracy=1
[38] Train-accuracy=1
[39] Train-accuracy=1
[40] Train-accuracy=1
[41] Train-accuracy=1
[42] Train-accuracy=1
[43] Train-accuracy=1
[44] Train-accuracy=1
[45] Train-accuracy=1
[46] Train-accuracy=1
[47] Train-accuracy=1
[48] Train-accuracy=1
[49] Train-accuracy=1
[50] Train-accuracy=1
[1]    10 10000
pred.label
    0    1    2    3    4    5    6    7    8    9
  992 1139 1029 1017  983  877  953 1021  972 1017
Accuracy (PCC): 98.21%
Cohen's Kappa: 0.9801
Users accuracy:
    0    1    2    3    4    5    6    7    8    9
 99.3 99.5 98.2 98.2 98.1 97.1 98.0 97.7 98.0 97.8
Producers accuracy:
    0    1    2    3    4    5    6    7    8    9
 98.1 99.1 98.4 97.5 98.0 98.7 98.5 98.3 98.3 97.1
```

```
Confusion matrix
     y
x        0      1      2      3      4      5      6      7      8      9
    0  973      0      2      2      1      3      5      1      3      2
    1    1   1129      0      0      1      1      3      2      0      2
    2    1      0   1013      1      3      0      0      9      2      0
    3    0      1      5    992      0     10      1      1      3      4
    4    0      0      2      0    963      2      7      1      1      7
    5    0      0      0      4      1    866      2      0      2      2
    6    2      2      1      0      3      5    939      0      1      0
    7    0      1      6      3      1      1      0   1004      2      3
    8    1      1      3      4      0      2      1      3    955      2
    9    2      1      0      4      9      2      0      7      5    987
```

We can observe from the output that 100% accuracy was obtained with the training dataset. However, with the test dataset, we observe the accuracy as 98%. Essentially, our model is expected to perform the same with both the training and test dataset for it to be called a good model. Unfortunately, in this case, we have encountered a situation known as **overfitting,** which means that the model we created did not generalize well. In other words, the model has trained itself with too many parameters or it got trained for too long and has become super-specialized with data in the training dataset alone; as an effect, it is not doing a good job with new data. Model generalization is something we should specifically aim for. There is a technique, known as **dropout**, that can help us to overcome the overfitting issue.

Implementing dropout to avoid overfitting

Dropout is defined in the network architecture after the activation layers, and it randomly sets activations to zero. In other words, dropout randomly deletes parts of the neural network, which allows us to prevent overfitting. We can't overfit exactly to our training data when we're consistently throwing away information learned along the way. This allows our neural network to learn to generalize better.

In MXNet, dropout can be easily defined as part of network architecture using the `mx.symbol.Dropout` function. For example, the following code defines dropouts post the first ReLU activation (`act1`) and second ReLU activation (`act2`):

```
dropout1 <- mx.symbol.Dropout(data = act1, p = 0.5)
dropout2 <- mx.symbol.Dropout(data = act2, p = 0.3)
```

The data parameter specifies the input that the dropout takes and the value of p specifies the amount of dropout to be done. In case of dropout1, we are specifying that 50% of weights are to be dropped. Again, there is no hard-and-fast rule in terms of how much dropout should be included and at what layers. This is something to be determined through trial and error. The code with dropouts almost remains identical to the earlier project except that it now includes the dropouts after the activations:

```
# code to read the dataset and transform it to train.x and train.y remains
# same as earlier project, therefore that code is not shown here
# including the required mxnet library
library(mxnet)
# defining the input layer in the network architecture
data <- mx.symbol.Variable("data")
# defining the first hidden layer with 128 neurons and also naming the #
layer as fc1
# passing the input data layer as input to the fc1 layer
fc1 <- mx.symbol.FullyConnected(data, name="fc1", num_hidden=128)
# defining the ReLU activation function on the fc1 output and also naming
the layer as ReLU1
act1 <- mx.symbol.Activation(fc1, name="ReLU1", act_type="relu")
# defining a 50% dropout of weights learnt
dropout1 <- mx.symbol.Dropout(data = act1, p = 0.5)
# defining the second hidden layer with 64 neurons and also naming the
layer as fc2
# passing the previous dropout output as input to the fc2 layer
fc2 <- mx.symbol.FullyConnected(dropout1, name="fc2", num_hidden=64)
# defining the ReLU activation function on the fc2 output and also naming
the layer as ReLU2
act2 <- mx.symbol.Activation(fc2, name="ReLU2", act_type="relu")
# defining a dropout with 30% weight drop
dropout2 <- mx.symbol.Dropout(data = act2, p = 0.3)
# defining the third and final hidden layer in our network with 10 neurons
and also naming the layer as fc3
# passing the previous dropout output as input to the fc3 layer
fc3 <- mx.symbol.FullyConnected(dropout2, name="fc3", num_hidden=10)
# defining the output layer with softmax activation function to
obtain class probabilities
softmax <- mx.symbol.SoftmaxOutput(fc3, name="sm")
# defining that the experiment should run on cpu
devices <- mx.cpu()
# setting the seed for the experiment so as to ensure that the results are
reproducible
mx.set.seed(0)
# building the model with the network architecture defined above
model <- mx.model.FeedForward.create(softmax, X=train.x, y=train.y,
ctx=devices, num.round=50, array.batch.size=100,array.layout = "rowmajor",
learning.rate=0.07, momentum=0.9,  eval.metric=mx.metric.accuracy,
```

```
initializer=mx.init.uniform(0.07),
epoch.end.callback=mx.callback.log.train.metric(100))
# making predictions on the test dataset
preds <- predict(model, test)
# verifying the predicted output
print(dim(preds))
# getting the label for each observation in test dataset; the predicted
class is the one with highest probability
pred.label <- max.col(t(preds)) - 1
# observing the distribution of predicted labels in the test
dataset
print(table(pred.label))
# including the rfUtilities library so as to use accuracy function
library(rfUtilities)
# obtaining the performance of the model
print(accuracy(pred.label,test.y))
# printing the network architecture
graph.viz(model$symbol)
```

This will give the following output and the visual network architecture:

```
[35] Train-accuracy=0.958950003186862
[36] Train-accuracy=0.958983335793018
[37] Train-accuracy=0.958083337446054
[38] Train-accuracy=0.959683336317539
[39] Train-accuracy=0.95990000406901
[40] Train-accuracy=0.959433337251345
[41] Train-accuracy=0.959066670437654
[42] Train-accuracy=0.960250004529953
[43] Train-accuracy=0.959983337720235
[44] Train-accuracy=0.960450003842513
[45] Train-accuracy=0.960150004227956
[46] Train-accuracy=0.960533337096373
[47] Train-accuracy=0.962033336758614
[48] Train-accuracy=0.96005000303189
[49] Train-accuracy=0.961366670827071
[50] Train-accuracy=0.961350003282229
[1]    10 10000
pred.label
    0     1     2     3     4     5     6     7     8     9
  984  1143  1042  1022   996   902   954  1042   936   979
Accuracy (PCC): 97.3%
Cohen's Kappa: 0.97
Users accuracy:
    0     1     2     3     4     5     6     7     8     9
 98.7  98.9  98.1  97.6  98.2  97.3  97.6  97.4  94.3  94.7
Producers accuracy:
    0     1     2     3     4     5     6     7     8     9
```

```
98.3 98.3 97.1 96.5 96.8 96.2 98.0 96.1 98.1 97.7
Confusion matrix
```

y	0	1	2	3	4	5	6	7	8	9
x										
0	967	0	0	0	0	2	5	1	6	3
1	0	1123	3	0	1	1	3	5	2	5
2	1	2	1012	4	3	0	0	14	4	2
3	2	1	4	986	0	6	1	3	12	7
4	0	0	3	0	964	2	5	0	5	17
5	2	3	0	9	0	868	7	0	9	4
6	3	2	0	0	5	3	935	0	6	0
7	4	1	9	4	3	3	0	1001	6	11
8	1	3	1	2	1	3	2	1	918	4
9	0	0	0	5	5	4	0	3	6	956

Take a look at the following diagram:

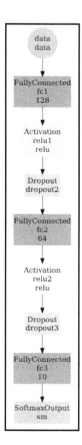

We can see from the output that dropout is now included as part of the network architecture. We also observe that this network architecture yields a lower accuracy on the test dataset when compared with our initial project. One reason could be that the dropout percentages (50% and 30%) we included are too high. We could play with these percentages and rebuild the model to determine whether the accuracy gets better. The idea, however, is to demonstrate the use of dropout as a regularization technique so as to avoid overfitting in deep neural networks.

Apart from dropout, there are other techniques you could employ to avoid an overfitting situation:

- **Addition of data**: Adding more training data.
- **Data augmentation**: Creating additional data synthetically by applying techniques such as flipping, distorting, adding random noise, and rotation. The following screenshot shows sample images created after applying data augmentation:

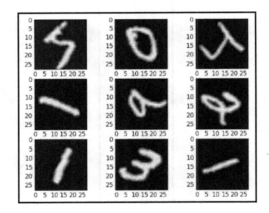

Sample images from applying data augmentation

- **Reducing complexity of the network architecture**: Fewer layers, fewer epochs, and so on.
- **Batch normalization**: A process of ensuring that the weights generated in the network do not push very high or very low. This is generally achieved by subtracting the mean and dividing by the standard deviation of all weights at a layer from each weight in a layer. It shields against overfitting, performs regularization, and significantly improves the training speed. The `mx.sym.batchnorm()` function enables us to define batch normalization after the activation.

We will not focus on developing another project with batch normalization as using this function in the project is very similar to the other functions we used in our earlier projects. So far, we have focused on increasing the epochs to improve the performance of the model, another option is to try a different architecture and evaluate whether that improves the accuracy on the test dataset. On that note, let's explore LeNet, which is specifically designed for optical character recognition in documents.

Implementing the LeNet architecture with the MXNet library

In their 1998 paper, *Gradient-Based Learning Applied to Document Recognition*, LeCun et al. introduced the LeNet architecture.

The LeNet architecture consists of two sets of convolutional, activation, and pooling layers, followed by a fully-connected layer, activation, another fully-connected layer, and finally a softmax classifier. The following diagram illustrates the LeNet architecture:

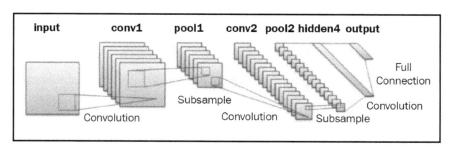

LeNet architecture

Now, let's implement the LeNet architecture with the mxnet library in our project using the following code block:

```
## setting the working directory
setwd('/home/sunil/Desktop/book/chapter 6/MNIST')
# function to load image files
load_image_file = function(filename) {
  ret = list()
  f = file(filename, 'rb')
  readBin(f, 'integer', n = 1, size = 4, endian = 'big')
  n    = readBin(f, 'integer', n = 1, size = 4, endian = 'big')
  nrow = readBin(f, 'integer', n = 1, size = 4, endian = 'big')
  ncol = readBin(f, 'integer', n = 1, size = 4, endian = 'big')
  x = readBin(f, 'integer', n = n * nrow * ncol, size = 1, signed
```

```
= FALSE)
  close(f)
  data.frame(matrix(x, ncol = nrow * ncol, byrow = TRUE))
}
# function to load label files
load_label_file = function(filename) {
  f = file(filename, 'rb')
  readBin(f, 'integer', n = 1, size = 4, endian = 'big')
  n = readBin(f, 'integer', n = 1, size = 4, endian = 'big')
  y = readBin(f, 'integer', n = n, size = 1, signed = FALSE)
  close(f)
  y
}
# load images
train = load_image_file("train-images-idx3-ubyte")
test  = load_image_file("t10k-images-idx3-ubyte")
# converting the train and test data into a format as required by LeNet
train.x <- t(data.matrix(train))
test <- t(data.matrix(test))
# loading the labels
train.y = load_label_file("train-labels-idx1-ubyte")
test.y  = load_label_file("t10k-labels-idx1-ubyte")
# linearly transforming the grey scale image i.e. between 0 and 255 to # 0
and 1
train.x <- train.x/255
test <- test/255
# including the required mxnet library
library(mxnet)
# input
data <- mx.symbol.Variable('data')
# first convolution layer
conv1 <- mx.symbol.Convolution(data=data, kernel=c(5,5), num_filter=20)
# applying the tanh activation function
tanh1 <- mx.symbol.Activation(data=conv1, act_type="tanh")
# applying max pooling
pool1 <- mx.symbol.Pooling(data=tanh1, pool_type="max", kernel=c(2,2),
stride=c(2,2))
# second conv
conv2 <- mx.symbol.Convolution(data=pool1, kernel=c(5,5), num_filter=50)
# applying the tanh activation function again
tanh2 <- mx.symbol.Activation(data=conv2, act_type="tanh")
#performing max pooling again
pool2 <- mx.symbol.Pooling(data=tanh2, pool_type="max",
                           kernel=c(2,2), stride=c(2,2))
# flattening the data
flatten <- mx.symbol.Flatten(data=pool2)
# first fullconnected later
fc1 <- mx.symbol.FullyConnected(data=flatten, num_hidden=500)
```

```
# applying the tanh activation function
tanh3 <- mx.symbol.Activation(data=fc1, act_type="tanh")
# second fullconnected layer
fc2 <- mx.symbol.FullyConnected(data=tanh3, num_hidden=10)
# defining the output layer with softmax activation function to obtain #
class probabilities
lenet <- mx.symbol.SoftmaxOutput(data=fc2)
# transforming the train and test dataset into a format required by
# MxNet functions
train.array <- train.x
dim(train.array) <- c(28, 28, 1, ncol(train.x))
test.array <- test
dim(test.array) <- c(28, 28, 1, ncol(test))
# setting the seed for the experiment so as to ensure that the
# results are reproducible
mx.set.seed(0)
# defining that the experiment should run on cpu
devices <- mx.cpu()
# building the model with the network architecture defined above
model <- mx.model.FeedForward.create(lenet, X=train.array, y=train.y,
ctx=devices, num.round=3, array.batch.size=100, learning.rate=0.05,
momentum=0.9, wd=0.00001, eval.metric=mx.metric.accuracy,
             epoch.end.callback=mx.callback.log.train.metric(100))
# making predictions on the test dataset
preds <- predict(model, test.array)
# getting the label for each observation in test dataset; the
# predicted class is the one with highest probability
pred.label <- max.col(t(preds)) - 1
# including the rfUtilities library so as to use accuracy
function
library(rfUtilities)
# obtaining the performance of the model
print(accuracy(pred.label,test.y))
# printing the network architecture
graph.viz(model$symbol,direction="LR")
```

This will give the following output and the visual network architecture:

```
Start training with 1 devices
[1] Train-accuracy=0.678916669438283
[2] Train-accuracy=0.978666676680247
[3] Train-accuracy=0.98676667680343
Accuracy (PCC): 98.54%
Cohen's Kappa: 0.9838
Users accuracy:
    0     1     2     3     4     5     6     7     8     9
 99.8 100.0  97.0  98.4  98.9  98.2  98.2  98.7  98.2  97.8
Producers accuracy:
```

```
     0    1    2    3    4    5    6    7    8    9
  98.0 96.9 99.1 99.3 99.0 99.3 99.6 97.7 98.7 98.3
  Confusion matrix
    y
  x      0    1    2    3    4    5    6    7    8    9
    0  978    0    2    2    1    3    7    0    4    1
    1    0 1135   15    2    1    0    5    7    1    5
    2    0    0 1001    2    1    1    0    3    2    0
    3    0    0    0  994    0    5    0    1    1    0
    4    0    0    1    0  971    0    1    0    0    8
    5    0    0    0    3    0  876    2    0    1    0
    6    0    0    0    0    2    1  941    0    1    0
    7    1    0    7    1    3    1    0 1015    3    8
    8    1    0    6    1    1    1    2    1  956    0
    9    0    0    0    5    2    4    0    1    5  987
```

Take a look at the following diagram:

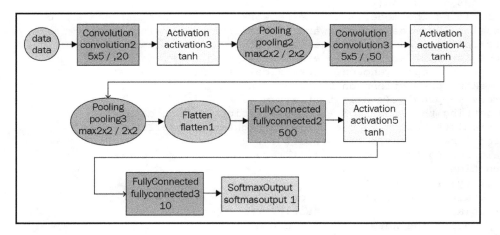

The code ran for less than 5 minutes on my 4-core CPU box, but still got us a 98% accuracy on the test dataset with just three epochs. We can also see that we obtained 98% accuracy with both the training and test datasets, confirming that there is no overfitting.

We see `tanh` is used as the activation function; let's experiment and see whether it has any impact if we change it to ReLU. The code for the project will be identical except that we need to find and replace `tanh` with ReLU. We will not repeat the code as the only lines that have changed from the earlier project are as follows:

```
ReLU1 <- mx.symbol.Activation(data=conv1, act_type="relu")
pool1 <- mx.symbol.Pooling(data=ReLU1, pool_type="max",
                           kernel=c(2,2), stride=c(2,2))
```

```
ReLU2 <- mx.symbol.Activation(data=conv1, act_type="relu")
pool2 <- mx.symbol.Pooling(data=ReLU2, pool_type="max",
                           kernel=c(2,2), stride=c(2,2))
ReLU3 <- mx.symbol.Activation(data=conv1, act_type="relu")
fc2 <- mx.symbol.FullyConnected(data=ReLU3, num_hidden=10)
```

You will get the following output on running the code with ReLU as the activation function:

```
Start training with 1 devices
[1] Train-accuracy=0.627283334874858
[2] Train-accuracy=0.979916676084201
[3] Train-accuracy=0.987366676231225
Accuracy (PCC): 98.36%
Cohen's Kappa: 0.9818
Users accuracy:
    0    1    2    3    4    5    6    7    8    9
 99.8 99.7 97.9 99.4 98.6 96.5 97.7 98.2 97.4 97.9
Producers accuracy:
    0    1    2    3    4    5    6    7    8    9
 97.5 97.2 99.6 95.6 99.7 99.2 99.7 98.0 99.6 98.2
Confusion matrix
   y
x       0    1    2    3    4    5    6    7    8    9
  0   978    0    3    1    1    2   12    0    5    1
  1     1 1132    6    0    2    1    5   11    1    6
  2     0    0 1010    1    0    0    0    1    2    0
  3     0    2    4 1004    0   23    1    3    9    4
  4     0    0    1    0  968    0    1    0    0    1
  5     0    1    0    1    0  861    2    0    3    0
  6     0    0    0    0    0    3  936    0    0    0
  7     1    0    6    3    0    1    0 1010    1    9
  8     0    0    2    0    1    0    1    0  949    0
  9     0    0    0    0   10    1    0    3    4  988
```

With ReLU being used as the activation function, we do not see a significant improvement in the accuracy. It stayed at 98%, which is the same as obtained with the tanh activation function.

As a next step, we could try to rebuild the model with additional epochs to see whether the accuracy improves. Alternatively, we could try tweaking the number of filters and filter sizes per convolutional layer to see what happens! Further experiments could also include adding more layers of several kinds. We don't know what the result is going to be unless we experiment!

Implementing computer vision with pretrained models

In Chapter 1, *Exploring the Machine Learning Landscape,* we touched upon a concept called **transfer learning**. The idea is to take the knowledge learned in a model and apply it to another related task. Transfer learning is used on almost all computer vision tasks nowadays. It's rare to train models from scratch unless there is a huge labeled dataset available for training.

Generally, in computer vision, CNNs try to detect edges in the earlier layers, shapes in the middle layer, and some task-specific features in the later layers. Irrespective of the image to be detected by the CNNs, the function of the earlier and middle layers remains the same, which makes it possible to exploit the knowledge gained by a pretrained model. With transfer learning, we can reuse the early and middle layers and only retrain the later layers. It helps us to leverage the labeled data of the task it was initially trained on.

Transfer learning offers two main advantages: it saves us training time and ensures that we have a good model even if we have a lot less labelled training data.

Xception, VGG16, VGG19, ResNet50, InceptionV3, InceptionResNetV2, MobileNet, DenseNet, NASNet, MobileNetV2, QuocNet, AlexNet, Inception (**GoogLeNet**), and BN-Inception-v2 are some widely-used pretrained models. While we won't delve into the details of each of these pretrained models, the idea of this section is to implement a project to detect the contents of images (input) by making use of a pretrained model through MXNet.

In the code presented in this section, we make use of the pretrained Inception-BatchNorm network to predict the class of an image. The pretrained model needs to be downloaded to the working directory prior to running the code. The model can be downloaded from http://data.mxnet.io/mxnet/data/Inception.zip. Let's explore the following code to label a few test images using the inception_bn pretrained model:

```
# loading the required libraries
library(mxnet)
library(imager)
# loading the inception_bn model to memory
model = mx.model.load("/home/sunil/Desktop/book/chapter
6/Inception/Inception_BN", iteration=39)
# loading the mean image
mean.img = as.array(mx.nd.load("/home/sunil/Desktop/book/chapter
6/Inception/mean_224.nd")[["mean_img"]])
# loading the image that need to be classified
im <- load.image("/home/sunil/Desktop/book/chapter 6/image1.jpeg")
```

```
# displaying the image
plot(im)
```

This will result in the following output:

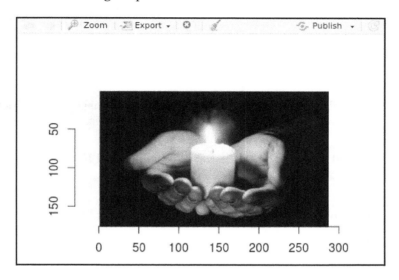

To process the images and predict the image IDs that have the highest probability of using the pretrained model, we use the following code:

```
# function to pre-process the image so as to be consumed by predict
function that is using inception_bn model
preproc.image <- function(im, mean.image) {
  # crop the image
  shape <- dim(im)
  short.edge <- min(shape[1:2])
  xx <- floor((shape[1] - short.edge) / 2)
  yy <- floor((shape[2] - short.edge) / 2)
  cropped <- crop.borders(im, xx, yy)
  # resize to 224 x 224, needed by input of the model.
  resized <- resize(cropped, 224, 224)
  # convert to array (x, y, channel)
  arr <- as.array(resized) * 255
  dim(arr) <- c(224, 224, 3)
  # subtract the mean
  normed <- arr - mean.img
  # Reshape to format needed by mxnet (width, height, channel,
num)
  dim(normed) <- c(224, 224, 3, 1)
  return(normed)
}
```

```
# calling the image pre-processing function on the image to be classified
normed <- preproc.image(im, mean.img)
# predicting the probabilties of labels for the image using the pre-trained
model
prob <- predict(model, X=normed)
# sorting and filtering the top three labels with highest
probabilities
max.idx <- order(prob[,1], decreasing = TRUE)[1:3]
# printing the ids with highest probabilities
print(max.idx)
```

This will result in the following output with the IDs of the highest probabilities:

```
[1] 471 627 863
```

Let's print the labels that correspond to the top-three predicted IDs with the highest probabilities using the following code:

```
# loading the pre-trained labels from inception_bn model
synsets <- readLines("/home/sunil/Desktop/book/chapter
6/Inception/synset.txt")
# printing the english labels corresponding to the top 3 ids with highest
probabilities
print(paste0("Predicted Top-classes: ", synsets[max.idx]))
```

This will give the following output:

```
[1] "Predicted Top-classes: n02948072 candle, taper, wax light"
[2] "Predicted Top-classes: n03666591 lighter, light, igniter, ignitor"
[3] "Predicted Top-classes: n04456115 torch"
```

From the output, we see that it has correctly labelled the image that is passed as input. We can test a few more images with the following code to confirm that the classification is done correctly:

```
im2 <- load.image("/home/sunil/Desktop/book/chapter 6/image2.jpeg")
plot(im2)
normed <- preproc.image(im2, mean.img)
prob <- predict(model, X=normed)
max.idx <- order(prob[,1], decreasing = TRUE)[1:3]
print(paste0("Predicted Top-classes: ", synsets[max.idx]))
```

This will give the following output:

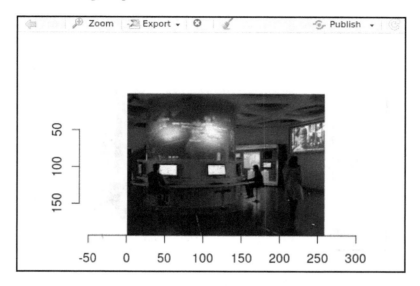

Take a look at the following code:

```
[1] "Predicted Top-classes: n03529860 home theater, home theatre"
[2] "Predicted Top-classes: n03290653 entertainment center"          [3]
"Predicted Top-classes: n04404412 television, television system"
```

Likewise, we can try for a third image using the following code:

```
# getting the labels for third image
im3 <- load.image("/home/sunil/Desktop/book/chapter
6/image3.jpeg")
plot(im3)
normed <- preproc.image(im3, mean.img)
prob <- predict(model, X=normed)
max.idx <- order(prob[,1], decreasing = TRUE)[1:3]
print(paste0("Predicted Top-classes: ", synsets[max.idx]))
```

This will give the following output:

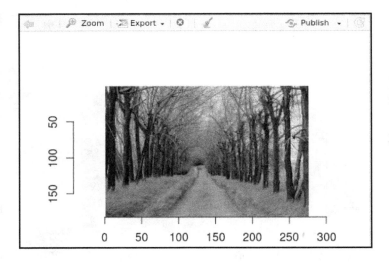

Take a look at the following output:

```
[1] "Predicted Top-classes: n04326547 stone wall"
[2] "Predicted Top-classes: n03891251 park bench"
[3] "Predicted Top-classes: n04604644 worm fence, snake fence, snake-rail
fence, Virginia fence"
```

Summary

In this chapter, we learned about computer vision and its association with deep learning. We explored a specific type of deep learning algorithm, CNNs, that is widely used in computer vision. We studied an open source deep learning framework called MXNet. After a detailed discussion of the MNIST dataset, we built models using various network architectures and successfully classified the handwritten digits in the MNIST dataset. At the end of the chapter, we delved into the concept of transfer learning and explored its association with computer vision. The last project we built in this chapter classified images using an Inception-BatchNorm pretrained model.

In the next chapter, we will explore an unsupervised learning algorithm called the autoencoder neural network. I am really excited to implement a project to capture credit card fraud using autoencoders. Are you game? Let's go!

7
Credit Card Fraud Detection Using Autoencoders

Fraud management has been known to be a very painful problem for banking and finance firms. Card-related frauds have proven to be especially difficult for firms to combat. Technologies such as chip and PIN are available and are already used by most credit card system vendors, such as Visa and MasterCard. However, the available technology is unable to curtail 100% of credit card fraud. Unfortunately, scammers come up with newer ways of phishing to obtain passwords from credit card users. Also, devices such as skimmers make stealing credit card data a cake walk!

Despite the availability of some technical abilities to combat credit card fraud, *The Nilson Report*, a leading publication covering payment systems worldwide, estimated that credit card fraud is going to soar to $32 billion in 2020 (`https://nilsonreport.com/upload/content_promo/The_Nilson_Report_10-17-2017.pdf`). To get a perspective on the estimated loss, it is more than the recent profits posted by companies such as Coca-Cola ($2 billion), Warren Buffet's Berkshire Hathaway ($24 billion), and JP Morgan Chase ($23.5 billion)!

While credit card chip technology-providing companies have been investing hugely to advance the technology to counter credit card fraud, in this chapter, we are going to examine whether and how far machine learning can help deal with the credit card fraud problem. We will cover the following topics as we progress through this chapter:

- Machine learning in credit card fraud detection
- Autoencoders and the various types
- The credit card fraud dataset
- Building AEs with the H2O library in R
- Implementation of auto encoder for credit card fraud detection

Machine learning in credit card fraud detection

The task of fraud detection often boils down to outlier detection, in which a dataset is verified to find potential anomalies in the data. Traditionally, this task was deemed a manual task, where risk experts checked all transactions manually. Even though there is a technical layer, it is purely based on a rules base that scans through each transaction, and then those shortlisted as suspicious are sent through for a manual review to make a final decision on the transaction. However, there are some major drawbacks to this system:

- Organizations need substantial fraud management budgets for manual review staff.
- Extensive training is required to train the employees working as manual review staff.
- Training the personnel to manually review transactions is time consuming and expensive.
- Even the most highly trained manual review staff carry certain biases, therefore making the whole review system inaccurate.
- Manual reviews increase the time required to fulfill a transaction. The customers might get frustrated with the long wait times required to pass a credit card transaction. This may impact the loyalty of customers.
- Manual reviews may yield false positives. A false positive not only affects the sale in the process but also lifetime value generated from the customer.

Fortunately, with the rise of **machine learning (ML)**, **artificial intelligence (AI)**, and deep learning, it became feasible to automate the manual credit card transaction review process to a large extent. This not only saves an intensive amount of labor but also yields better detection of credit card fraud, which otherwise is impacted due to biases that human reviewers carry.

ML-based fraud detection strategies generally can be accomplished using both supervised ML and unsupervised ML techniques.

Supervised ML models are generally used when large amounts of transaction data tagged as **genuine** or **fraud** are available. A model is trained on the labeled dataset and the resultant model is then used for classifying any new credit card transactions into one of the two possible classes.

With most organizations, the problem is that labeled data is unavailable, or very little labeled data is available. This makes supervised learning models less feasible. This is where unsupervised models come into play. They are designed to spot anomalous behavior in transactions and they do not need explicit pre-labeled data to identify the anomalous behavior. The general idea in unsupervised fraud detection is to detect behavior anomalies by identifying transactions that do not conform to the majority.

Another thing to keep in mind is that fraud events are rare, and are not as common as genuine transactions. Due to the rarity of fraud, severe class imbalance problem may be seen in datasets related to credit card fraud. In other words, one would observe that 95% or more of the data in the dataset is of genuine transactions, and less than 5% of the data belongs to fraudulent transactions. Also, even if you learn about a fraudulent transaction today, the model is likely to face an anomaly tomorrow with different features. So, the problem space of genuine transactions is well known and it is pretty much stagnant; however, the problem space for fraudulent transactions is not well known and it is not constant. Due to these reasons, it make sense to deal with the fraud detection problem with unsupervised learning rather than supervised learning.

Anomaly detection is an unsupervised learning algorithm that is also termed a **one-class classification** algorithm. It distinguishes between **normal** and **anomalous** observations. The key principle on which the algorithm is built is that anomalous observations do not conform to the expected pattern of other common observations in a dataset. It is called a one-class classification as it learns the pattern of genuine transactions, and anything that shows non-conformance to this pattern is termed as an **anomaly**, and therefore as a fraudulent **transaction**. The following figure is an illustration showing anomaly detection in a two-dimensional space:

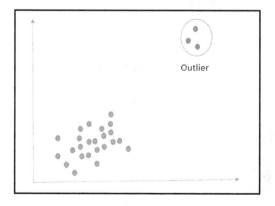

Anomaly detection illustrated in 2D space

A simple example of an anomaly is the identification of data points that are too far from the mean (standard deviation) in a time series. The following figure is an illustration displaying the data points that are identified as anomalies in a time series:

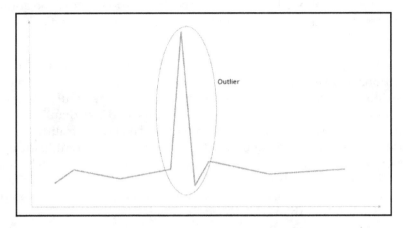

Anomaly in time series—identified through standard deviation

In this chapter, we will focus our efforts on a type of unsupervised deep learning application known as **AEs**.

Autoencoders explained

Autoencoders (**AEs**) are neural networks that are of a feedforward and non-recurrent type. They aim to copy the given inputs to the outputs. An AE works by compressing the input into a lower dimensional summary. This summary is often referred as latent space representation. An AE attempts to reconstruct the output from the latent space representation. An **Encoder**, a **Latent Space Representation**, and a **Decoder** are the three parts that make up the AEs. The following figure is an illustration showing the application of an AE on a sample picked from the MNIST dataset:

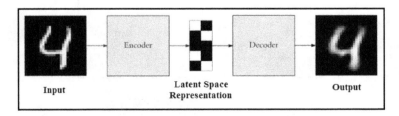

Application of AE on MNIST dataset sample

The encoder and decoder components of AEs are fully-connected feedforward networks. The number of neurons in a latent space representation is a hyperparameter that needs to be passed as part of building the AE. The number of neurons or nodes that is decided in the latent semantic space dictates the amount of compression that is attained while compressing the actual input image into a latent space representation. The general architecture of an AE is shown in the following figure:

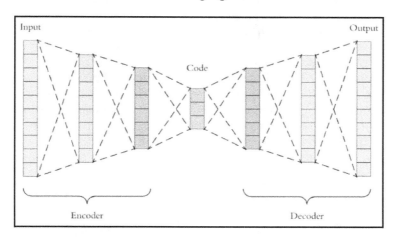

General architecture of a AE

The given input first passes through an **Encoder**, which is a fully-connected **artificial neural network (ANN)**. The **Encoder** acts upon the **Input** and reduces its dimensions, as specified in the hyperparameter. The **Decoder** is another fully-connected ANN that picks up this reduced **Input** (latent space representation) and then reconstructs the **Output**. The goal is to get the **Output** identical to that of the **Input**. In general, the architectures of the **Encoder** and the **Decoder** are mirror images. Although there is no such requirement that mandates that the **Encoder** and **Decoder** architectures should be the same, it is generally practiced that way. In fact, the only requirement of the AE is to obtain identical output from that of the given input. Anything in between can be customized to the whims and fancies of the individual building the AE.

Mathematically, the encoder can be represented as:

$$y = h(x)$$

where x is the input and h is the function that acts on the input to represent it in a concise summary format. A decoder, on the other hand, can be represented as:

$$r = f(y) = f\big(h(x)\big)$$

While the expectation is to obtain $r = x$, this is not always the case as the reconstruction is done from a compact summary representation; therefore, there is occurrence of certain error. The error e is computed from the original input x and reconstructed output r, $e = x - r$.

The AE network then learns by reducing the **Mean Squared Error (MSE)**, and the error is propagated back to the hidden layers for adjustment. The weights of the decoder and encoder are transposes of each other, which makes it faster to learn training parameters. The mirrored architectures of the encoder and decoder make it possible to learn the training parameters faster. In different architectures, the weights cannot be simply transposed; therefore, the computation time will increase. This is the reason for keeping the mirrored architectures for the encoder and decoder.

Types of AEs based on hidden layers

Based on the size of the hidden layer, AEs can be classified into two types, **undercomplete AEs** and **overcomplete AEs**:

- **Undercomplete AE**: If the AE simply learns to copy the input to the output, then it is not useful. The idea is to produce a concise representation as the output of the encoder, and this concise representation should consist of the most useful features of the input. The amount of conciseness achieved by the input layer is governed by the number of neurons or nodes that we use in the latent space representation. This can be set as a parameter while building the AE. If the number of neurons is set to fewer dimensions than that of the input features, then the AE is forced to learn most of the key features of the input data. The architecture where the number of neurons in latent space is less than that of input dimensions is called an undercomplete AE.
- **Overcomplete AE**: It is possible to represent the number of neurons in latent space as equal to or more than that of the input dimensions. This kind of architecture is termed an overcomplete AE. In this case, the AE does not learn anything and simply copies the input to the latent space, which in turn is propagated through to the decoder.

Apart from the number of neurons in the latent space, the following are some of the other parameters that can be used in an AE architecture:

- **Number of layers in the encoder and decoder**: The depth of the encoder and decoder can be set to any number. Generally, in a mirrored architecture of encoder and decoder, the number of layers is set as the same number. The last figure is an illustration showing the AE with two layers, excluding the input and output, in both the encoder and decoder.

- **Number of neurons per layer in encoder and decoder**: The number of neurons decreases with each layer in an encoder and it increases with each layer in a decoder. The neurons in layers of encoders and decoders are symmetric.

- **Loss function**: Loss functions such as MSE or cross-entropy are used by AEs to learn the weights during backpropagation. If the input is in the range of (0,1), then cross-entropy is used as metric, otherwise MSE is used.

Types of AEs based on restrictions

Based on the restrictions imposed on the loss, AEs can be grouped into the following types:

- **Plain Vanilla AEs**: This is the simplest AE architecture possible, with a fully-connected neural layer as the encoder and decoder.

- **Sparse AEs**: Sparse AEs are an alternative method for introducing an information bottleneck, without requiring a reduction in the number of nodes in our hidden layers. Rather than preferring an undercomplete AE, the loss function is constructed in a way that it penalizes the activations within a layer. For any given observation, the network is encouraged to learn encoding and decoding, which only relies on activating a small number of neurons.

- **Denoising AEs**: This is a type of overcomplete AE that experiences the risk of learning the **identity function** or **null function**. Essentially, the AE learns the output that is equal to the input, therefore making the AE useless. Denoising AEs avoid this problem of learning the identity function by randomly initializing some of the inputs to 0. During the computation of the loss function, the noise-induced input is not considered; therefore, the network still learns the correct weights without the risk of learning the identity function. At the same time, the AE is trained to learn to reconstruct the output, even from the corrupted input.

The following figure is a example of denoising AEs on sample images from the MNIST dataset:

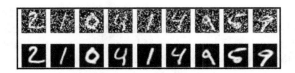

Application of denoising AEs on MNIST samples

- **Convolutional AEs**: When dealing with images as inputs, one can use convolutional layers as part of the encoder and decoder networks. Such kinds of AEs that use convolutional layers are termed **convolutional AEs**. The following figure is an illustration showing the use of convolutions in AEs:

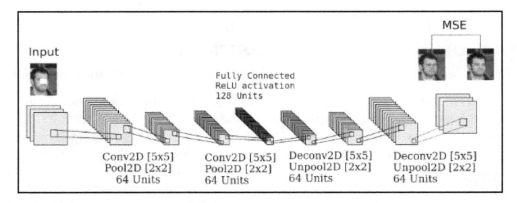

Convolutional AEs

- **Stacked AEs**: Stacked AEs are ones that have multiple layers in the encoder as well as the decoder. You can refer to the general architecture of an AE as an example illustration of a stacked AE architecture, with the encoder and decoder having two layers (excluding the input and output layers).
    - **Variational AEs**: A **variational AE** (**VAE**), rather than building an encoder that outputs a singl
- e value to describe each latent state attribute, describes a probability distribution for each latent attribute. This makes it possible to design complex generative models of data and also generate fictional celebrity images and digital artwork. The following figure is an illustration depicting the representation of data in VAEs:

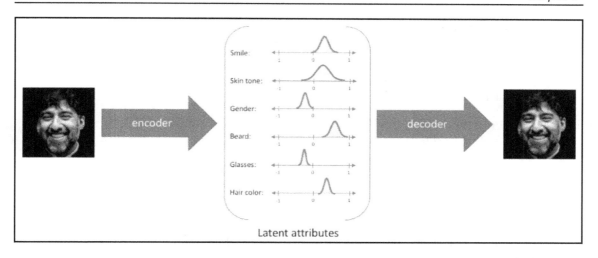

Latent attributes

In a VAE, the encoder model is sometimes referred to as the recognition model, whereas the decoder model is sometimes referred to as the generative model. The encoder outputs a range of statistical distributions for the latent features. These features are randomly sampled and used by the decoder to reconstruct the input. For any sampling of the latent distributions, the decoder is expected to be able to accurately reconstruct the input. Thus, values that are nearby to one another in latent space should correspond with very similar reconstructions.

Applications of AEs

The following are some of the practical applications where AEs may be used:

- **Image coloring**: Given a grayscale image as input, AEs can auto color the image and return the colored image as output.
- **Noise removal**: Denoising AEs are able to remove noise from images and reconstruct images without noise. Tasks such as watermark removal from videos and images can be accomplished.
- **Dimensionality reduction**: AEs represent the input data in a compressed form, but with a focus on key features alone. Therefore, things like images can be represented with reduced pixels, without much loss of information during image reconstruction.
- **Image search**: This is used to identify similar images based on a given input.

- **Information retrieval**: When retrieving information from a corpus, AEs may be used to group together all the documents that belong to a given input.
- **Topic modeling**: Variational AEs are used to approximate the posterior distribution, and it has become a promising alternative for inferring latent topic distributions of text documents.

We have covered the fundamentals that are needed for us to understand AEs and their applications. Let us understand, at a high level, the solution we are going to employ using AEs on the credit card fraud detection problem.

The credit card fraud dataset

Generally in a fraud dataset, we have sufficient data for the negative class (non-fraud/genuine transactions) and very few or no data for the positive class (fraudulent transactions). This is termed a **class imbalance problem** in the ML world. We train an AE on the non-fraud data and learn features using the encoder. The decoder is then used to compute the reconstruction error on the training set to find a threshold. This threshold will be used on the unseen data (test dataset or otherwise). We use the threshold to identify those test instances whose values are greater than the threshold as fraud instances.

For the project in this chapter, we will be using a dataset that is sourced from this URL: `https://essentials.togaware.com/data/`. This is a public dataset of credit card transactions. This dataset is originally made available through the research paper *Calibrating Probability with Undersampling for Unbalanced Classification*, A. Dal Pozzolo, O. Caelen, R. A Johnson and G. Bontempi, IEEE **Symposium Series on Computational Intelligence** (**SSCI**), Cape Town, South Africa, 2015. The dataset is also available at this URL: `http://www.ulb.ac.be/di/map/adalpozz/data/creditcard.Rdata`. The dataset was collected and analyzed during a research collaboration of Worldline and the Machine Learning Group (`http://mlg.ulb.ac.be`) of ULB (Université Libre de Bruxelles) on big data mining and fraud detection.

The following are the characteristics of the dataset:

- The paper made the dataset available as an Rdata file. There is a CSV converted version of this dataset available on Kaggle as well as other sites.
- It contains transactions made by credit cards in September 2013 by European cardholders.
- The transactions occurred on two days are recorded and is presented as the dataset.

- There are a total of 284,807 transactions in the dataset.
- The dataset suffers from a severe class imbalance problem. Only 0.172% of all transactions are fraudulent transactions (492 fraudulent transactions).
- There are a total thirty features in the dataset, namely V1, V2, ...,V28, Time, and Amount.
- The variables V1, V2, ...,V28 are the principal components obtained with PCA from the original set of variables.
- Due to confidentiality, the original set of variables that yielded the principal components are not revealed.
- The Time feature contains the seconds elapsed between each transaction and the first transaction in the dataset.
- The Amount feature is the transaction amount.
- The dependent variable is named Class. The fraudulent transactions are represented as 1 in the class and genuine transactions are represented as 0.

We will now jump into using AEs for the credit card fraud detection.

Building AEs with the H2O library in R

We will be using the AE implementation available in H2O for our project. H2O is a fully open source, distributed, in-memory ML platform with linear scalability. It offers parallelized implementations of some of the most widely used ML algorithms. It supports an easy to use, unsupervised, and non-linear AE as part of its deep learning model. The DL AE of H2O is based on the multilayer neural net architecture, where the entire network is trained together, instead of being stacked layer by layer.

The h2o package can be installed in R with the following command:

```
install.packages("h2o")
```

Additional details on the installation and dependencies of H2O in R are available at this URL: https://cran.r-project.org/web/packages/h2o/index.html.

Once the package is installed successfully, the functions offered by the h2o package, including the AE, can simply be used by including the following line in R code:

```
library(h2o)
```

This is all we need to do prior to coding our credit card fraud detection system with the AE. Without waiting any longer, let's start building our code to explore and prepare our dataset, as well as to implement the AE that captures fraudulent credit card transactions.

Autoencoder code implementation for credit card fraud detection

As usual, like all other projects, let's first load the data into an R dataframe and then perform EDA to understand the dataset better. Please note the inclusion of h2o as well as the doParallel library in the code. These inclusions enable us to use the AE that is part of the h2o library, as well as to utilize the multiple CPU cores that are present in the laptop/desktop as follows:

```
# including the required libraries
library(tidyverse)
library(h2o)
library(rio)
library(doParallel)
library(viridis)
library(RColorBrewer)
library(ggthemes)
library(knitr)
library(caret)
library(caretEnsemble)
library(plotly)
library(lime)
library(plotROC)
library(pROC)
```

Initializing the H2O cluster in localhost under the port 54321. The nthreads defines the number of thread pools to be used, this is close to the number of cpus to be used. In our case, we are saying use all CPUs, we are also specifying the maximum memory to use by H2O cluster as 8G:

```
localH2O = h2o.init(ip = 'localhost', port = 54321, nthreads =
-1,max_mem_size = "8G")
# Detecting the available number of cores
no_cores <- detectCores() - 1
# utilizing all available cores
cl<-makeCluster(no_cores)
registerDoParallel(cl)
```

You will get a similar output to that shown in the following code block:

```
H2O is not running yet, starting it now...
Note:  In case of errors look at the following log files:
    /tmp/RtmpKZvQ3m/h2o_sunil_started_from_r.out
    /tmp/RtmpKZvQ3m/h2o_sunil_started_from_r.err
java version "1.8.0_191"
Java(TM) SE Runtime Environment (build 1.8.0_191-b12)
Java HotSpot(TM) 64-Bit Server VM (build 25.191-b12, mixed mode)
Starting H2O JVM and connecting: ..... Connection successful!
R is connected to the H2O cluster:
    H2O cluster uptime:         4 seconds 583 milliseconds
    H2O cluster timezone:       Asia/Kolkata
    H2O data parsing timezone:  UTC
    H2O cluster version:        3.20.0.8
    H2O cluster version age:    2 months and 27 days
    H2O cluster name:           H2O_started_from_R_sunil_jgw200
    H2O cluster total nodes:    1
    H2O cluster total memory:   7.11 GB
    H2O cluster total cores:    4
    H2O cluster allowed cores:  4
    H2O cluster healthy:        TRUE
    H2O Connection ip:          localhost
    H2O Connection port:        54321
    H2O Connection proxy:       NA
    H2O Internal Security:      FALSE
    H2O API Extensions:         XGBoost, Algos, AutoML, Core V3, Core V4
    R Version:                  R version 3.5.1 (2018-07-02)
```

Now, to set the working directory of the data file location, load Rdata and read it into the dataframe, and view the dataframe using the following code:

```
# setting the working directory where the data file is location
setwd("/home/sunil/Desktop/book/chapter 7")
# loading the Rdata file and reading it into the dataframe called cc_fraud
cc_fraud<-get(load("creditcard.Rdata"))
# performing basic EDA on the dataset
# Viewing the dataframe to confirm successful load of the dataset
View(cc_fraud)
```

The will give the following output:

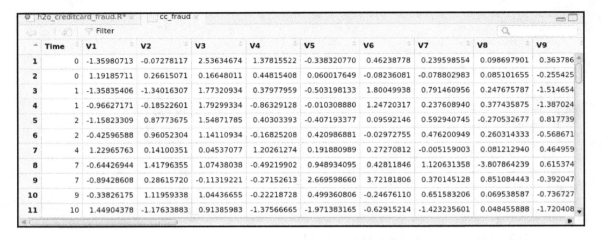

Let's now print the dataframe structure using the following code:

```
print(str(cc_fraud))
```

This will give the following output:

```
'data.frame':     284807 obs. of  31 variables:
$ Time    : num   0 0 1 1 2 2 4 7 7 9 ...
$ V1      : num   -1.36 1.192 -1.358 -0.966 -1.158 ...
$ V2      : num   -0.0728 0.2662 -1.3402 -0.1852 0.8777 ...
$ V3      : num   2.536 0.166 1.773 1.793 1.549 ...
$ V4      : num   1.378 0.448 0.38 -0.863 0.403 ...
$ V5      : num   -0.3383 0.06 -0.5032 -0.0103 -0.4072 ...
$ V6      : num   0.4624 -0.0824 1.8005 1.2472 0.0959 ...
$ V7      : num   0.2396 -0.0788 0.7915 0.2376 0.5929 ...
$ V8      : num   0.0987 0.0851 0.2477 0.3774 -0.2705 ...
$ V9      : num   0.364 -0.255 -1.515 -1.387 0.818 ...
$ V10     : num   0.0908 -0.167 0.2076 -0.055 0.7531 ...
$ V11     : num   -0.552 1.613 0.625 -0.226 -0.823 ...
$ V12     : num   -0.6178 1.0652 0.0661 0.1782 0.5382 ...
$ V13     : num   -0.991 0.489 0.717 0.508 1.346 ...
$ V14     : num   -0.311 -0.144 -0.166 -0.288 -1.12 ...
$ V15     : num   1.468 0.636 2.346 -0.631 0.175 ...
$ V16     : num   -0.47 0.464 -2.89 -1.06 -0.451 ...
$ V17     : num   0.208 -0.115 1.11 -0.684 -0.237 ...
$ V18     : num   0.0258 -0.1834 -0.1214 1.9658 -0.0382 ...
$ V19     : num   0.404 -0.146 -2.262 -1.233 0.803 ...
$ V20     : num   0.2514 -0.0691 0.525 -0.208 0.4085 ...
$ V21     : num   -0.01831 -0.22578 0.248 -0.1083 -0.00943 ...
```

```
$ V22    : num   0.27784 -0.63867 0.77168 0.00527 0.79828 ...
$ V23    : num   -0.11 0.101 0.909 -0.19 -0.137 ...
$ V24    : num   0.0669 -0.3398 -0.6893 -1.1756 0.1413 ...
$ V25    : num   0.129 0.167 -0.328 0.647 -0.206 ...
$ V26    : num   -0.189 0.126 -0.139 -0.222 0.502 ...
$ V27    : num   0.13356 -0.00898 -0.05535 0.06272 0.21942 ...
$ V28    : num   -0.0211 0.0147 -0.0598 0.0615 0.2152 ...
$ Amount: num   149.62 2.69 378.66 123.5 69.99 ...
$ Class : Factor w/ 2 levels "0","1": 1 1 1 1 1 1 1 1 1 1 ...
```

Now, to view the class distribution, use the following code:

```
print(table(cc_fraud$Class))
```

You will get the following output:

```
     0        1
284315      492
```

To view the relationship between the V1 and Class variables, use the following code:

```
# Printing the Histograms for Multivariate analysis
theme_set(theme_economist_white())
# visualization showing the relationship between variable V1 and the class
ggplot(cc_fraud,aes(x="",y=V1,fill=Class))+geom_boxplot()+labs(x="V1",y="")
```

This will give the following output:

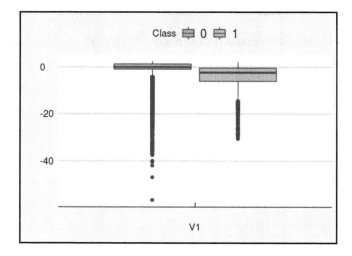

To visualize the distribution of transaction amounts with respect to class, use the following code:

```
# visualization showing the distribution of transaction amount with
# respect to the class, it may be observed that the amount are discretized
# into 50 bins for plotting purposes
ggplot(cc_fraud,aes(x = Amount)) + geom_histogram(color = "#D53E4F", fill =
"#D53E4F", bins = 50) + facet_wrap( ~ Class, scales = "free", ncol = 2)
```

This will give the following output:

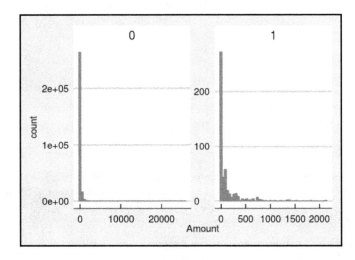

To visualize the distribution of transaction times with respect to class, use the following code:

```
ggplot(cc_fraud, aes(x =Time,fill = Class))+ geom_histogram(bins = 30)+
    facet_wrap( ~ Class, scales = "free", ncol = 2)
```

This will give the following output:

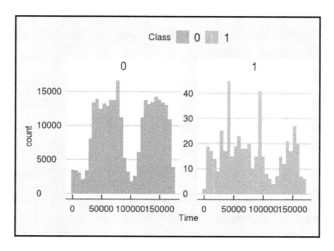

Use the following code to visualize the V2 variable with respect to Class:

```
ggplot(cc_fraud, aes(x =V2, fill=Class))+ geom_histogram(bins = 30)+
    facet_wrap( ~ Class, scales = "free", ncol = 2)
```

You will get the following as the output:

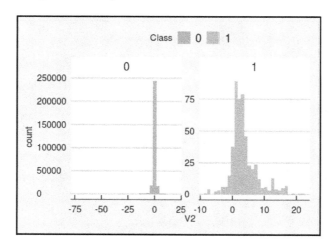

Use the following code to visualize V3 with respect to Class:

```
ggplot(cc_fraud, aes(x =V3, fill=Class))+ geom_histogram(bins = 30)+
    facet_wrap( ~ Class, scales = "free", ncol = 2)
```

The following graph is the resultant output:

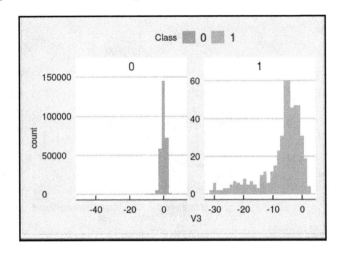

To visualize the V3 variable with respect to Class, use the following code:

```
ggplot(cc_fraud, aes(x =V4,fill=Class))+ geom_histogram(bins = 30)+
    facet_wrap( ~ Class, scales = "free", ncol = 2)
```

The following graph is the resultant output:

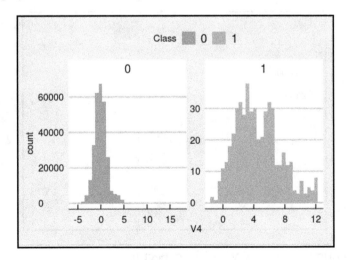

Use the following code to visualize the V6 variable with respect to `Class`:

```
ggplot(cc_fraud, aes(x=V6, fill=Class)) + geom_density(alpha=1/3) +
scale_fill_hue()
```

The following graph is the resultant output:

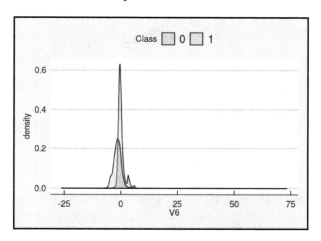

Use the following code to visualize the V7 variable with respect to `Class`:

```
ggplot(cc_fraud, aes(x=V7, fill=Class)) + geom_density(alpha=1/3) +
scale_fill_hue()
```

The following graph is the resultant output:

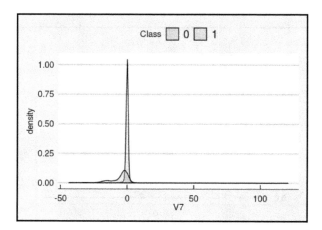

Use the following code to visualize the V8 variable with respect to Class:

```
ggplot(cc_fraud, aes(x=V8, fill=Class)) + geom_density(alpha=1/3) +
scale_fill_hue()
```

The following graph is the resultant output:

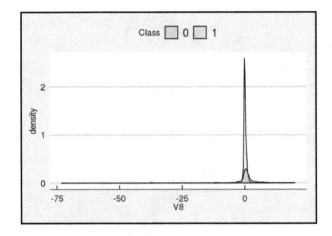

To visualize the V9 variable with respect to Class, use the following code:

```
# visualizationshowing the V7 variable with respect to the class
ggplot(cc_fraud, aes(x=V9, fill=Class)) + geom_density(alpha=1/3) +
scale_fill_hue()
```

The following graph is the resultant output:

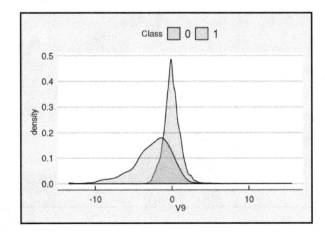

To visualize the `V10` variable with respect to `Class`, use the following code:

```
# observe we are plotting the data quantiles
ggplot(cc_fraud, aes(x ="",y=V10, fill=Class))+ geom_violin(adjust =
.5,draw_quantiles = c(0.25, 0.5, 0.75))+labs(x="V10",y="")
```

The following graph is the resultant output:

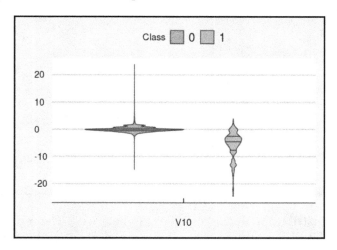

From all the visualizations related to variables with respect to class, we can infer that most of the principal components are centered on 0. Now, to plot the distribution of classes in the data, use the following code:

```
cc_fraud %>%
  ggplot(aes(x = Class)) +
  geom_bar(color = "chocolate", fill = "chocolate", width = 0.2) +
  theme_bw()
```

The following bar graph is the resultant output:

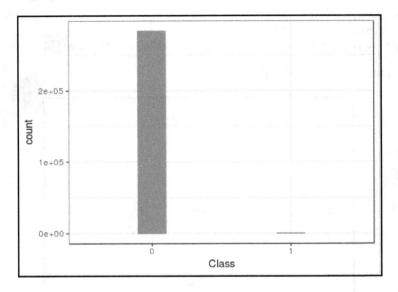

We observe that the distribution of classes is very imbalanced. The representation of the major class (non-fraudulent transactions, represented by 0) in the dataset is too heavy when compared to the minority class (fraudulent transactions: 1). In the traditional supervised ML way of dealing with this kind of problem, we would have treated the class imbalance problem with techniques such as **Synthetic Minority Over-Sampling Technique (SMORT)**. However, with AEs, we do not treat the class imbalance during data preprocessing; rather, we feed the data as is to the AE for learning. In fact, the AE is learning the thresholds and the characteristics of the data from the majority class; this is the reason we call it a one-class classification problem.

We will need to do some feature engineering prior to training our AE. Let's first focus on the Time variable in the data. Currently, it is in the seconds format, but we may better represent it as days. Run the following code to see the current form of time in the dataset:

```
print(summary(cc_fraud$Time))
```

You will get the following output:

```
Min. 1st Qu.  Median   Mean 3rd Qu.    Max.
   0   54202   84692  94814  139320  172792
```

We know that there are 86,400 seconds in a given day (60 seconds per minute * 60 minutes per hour * 24 hours per day). We will convert the `Time` variable into `Day` by considering the value in `Time` and representing it as `day1` if the number of seconds is less than or equal to 86,400, and anything over 86,400 becomes `day2`. There are only two days possible, as we can see from the summary that the maximum value represented by the time variable is `172792` seconds:

```
# creating a new variable called day based on the seconds
# represented in Time variable
 cc_fraud=cc_fraud %>% mutate(Day = case_when(.$Time > 3600 * 24 ~
"day2",.$Time < 3600 * 24 ~ "day1"))
#visualizing the dataset post creating the new variable
View(cc_fraud%>%head())
```

The following is the resultant output of the first six rows after the conversion:

21	V22	V23	V24	V25	V26	V27	V28	Amount	Class	Day
.018306778	0.277837576	-0.11047391	0.06692807	0.1285394	-0.1891148	0.133558377	-0.02105305	149.62	0	day1
.225775248	-0.638671953	0.10128802	-0.33984648	0.1671704	0.1258945	-0.008983099	0.01472417	2.69	0	day1
.247998153	0.771679402	0.90941226	-0.68928096	-0.3276418	-0.1390966	-0.055352794	-0.05975184	378.66	0	day1
.108300452	0.005273597	-0.19032052	-1.17557533	0.6473760	-0.2219288	0.062722849	0.06145763	123.50	0	day1
.009430697	0.798278495	-0.13745808	0.14126698	-0.2060096	0.5022922	0.219422230	0.21515315	69.99	0	day1
.208253515	-0.559824796	-0.02639767	-0.37142658	-0.2327938	0.1059148	0.253844225	0.08108026	3.67	0	day1

Now, use the following code to view the last six rows:

```
View(cc_fraud%>%tail())
```

The following is the resultant output of the last six rows after the conversion:

V21	V22	V23	V24	V25	V26	V27	V28	Amount	Class	Day
-0.3142046	-0.8085204	0.05034266	0.102799590	-0.4358701	0.1240789	0.217939865	0.06880333	2.69	0	day2
0.2134541	0.1118637	1.01447990	-0.509348453	1.4368069	0.2500343	0.943651172	0.82373096	0.77	0	day2
0.2142053	0.9243836	0.01246304	-1.016225669	-0.6066240	-0.3952551	0.068472470	-0.05352739	24.79	0	day2
0.2320450	0.5782290	-0.03750086	0.640133881	0.2657455	-0.0873706	0.004454772	-0.02656083	67.88	0	day2
0.2652449	0.8000487	-0.16329794	0.123205244	-0.5691589	0.5466685	0.108820735	0.10453282	10.00	0	day2
0.2610573	0.6430784	0.37677701	0.008797379	-0.4736487	-0.8182671	-0.002415309	0.01364891	217.00	0	day2

Now, let's print the distribution of transactions by the day in which the transaction falls, using the following code:

```
print(table(cc_fraud[,"Day"]))
```

You will get the following as the output:

```
  day1    day2
144786 140020
```

Let's create a new variable, `Time_day`, based on the seconds represented in the `Time` variable, and summarize the `Time_day` variable with respect to `Day` using the following code:

```
cc_fraud$Time_day <- if_else(cc_fraud$Day == "day2", cc_fraud$Time - 86400,
cc_fraud$Time)
print(tapply(cc_fraud$Time_day,cc_fraud$Day,summary,simplify = FALSE))
```

We get the following as the resultant output:

```
$day1
   Min. 1st Qu.  Median    Mean 3rd Qu.    Max.
      0   38432   54689   52948   70976   86398

$day2
   Min. 1st Qu.  Median    Mean 3rd Qu.    Max.
      1   37843   53425   51705   68182   86392
```

Use the following code the convert all character variables in the dataset to factors:

```
cc_fraud<-cc_fraud%>%mutate_if(is.character,as.factor)
```

We can further fine-tune the `Time_day` variable by converting the variable into a factor. The factors represents the time of day at which the transaction happened, for example, `morning`, `afternoon`, `evening`, and `night`. We can create a new variable called `Time_Group`, based on the various buckets of the day, using the following code:

```
cc_fraud=cc_fraud %>%
  mutate(Time_Group = case_when(.$Time_day <= 38138~ "morning" ,
                                .$Time_day <= 52327~  "afternoon",
                                .$Time_day <= 69580~"evening",
                                .$Time_day > 69580~"night"))
#Visualizing the data post creating the new variable
View(head(cc_fraud))
```

The following is the resultant output of the first six rows:

V24	V25	V26	V27	V28	Amount	Class	Day	Time_day	Time_Group
0.06692807	0.1285394	-0.1891148	0.133558377	-0.02105305	149.62	0	day1	0	morning
-0.33984648	0.1671704	0.1258945	-0.008983099	0.01472417	2.69	0	day1	0	morning
-0.68928096	-0.3276418	-0.1390966	-0.055352794	-0.05975184	378.66	0	day1	1	morning
-1.17557533	0.6473760	-0.2219288	0.062722849	0.06145763	123.50	0	day1	1	morning
0.14126698	-0.2060096	0.5022922	0.219422230	0.21515315	69.99	0	day1	2	morning
-0.37142658	-0.2327938	0.1059148	0.253844225	0.08108026	3.67	0	day1	2	morning

Use the following code to view and confirm the last six rows:

```
View(tail(cc_fraud))
```

This will give the following output, and we see that we have successfully converted the data which represent the various time of the day:

V24	V25	V26	V27	V28	Amount	Class	Day	Time_day	Time_Group
0.102799590	-0.4358701	0.1240789	0.217939865	0.06880333	2.69	0	day2	86385	night
-0.509348453	1.4368069	0.2500343	0.943651172	0.82373096	0.77	0	day2	86386	night
-1.016225669	-0.6066240	-0.3952551	0.068472470	-0.05352739	24.79	0	day2	86387	night
0.640133881	0.2657455	-0.0873706	0.004454772	-0.02656083	67.88	0	day2	86388	night
0.123205244	-0.5691589	0.5466685	0.108820735	0.10453282	10.00	0	day2	86388	night
0.008797379	-0.4736487	-0.8182671	-0.002415309	0.01364891	217.00	0	day2	86392	night

Take a look at the following code:

```
#visualizing the transaction count by day
cc_fraud %>%drop_na()%>%
  ggplot(aes(x = Day)) +
  geom_bar(fill = "chocolate",width = 0.3,color="chocolate") +
  theme_economist_white()
```

The preceding code will generate the following output:

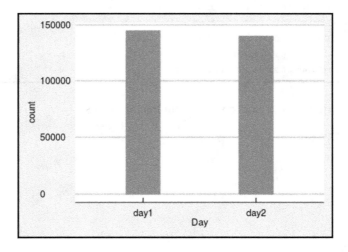

We can infer from the visualization that there is no difference in the count of transactions that happened on day 1 and day 2. Both remain close to 150,000 transactions.

Now we will convert the Class variable as a factor and then visualize the data by Time_Group variable using the following code:

```
cc_fraud$Class <- factor(cc_fraud$Class)
cc_fraud %>%drop_na()%>%
  ggplot(aes(x = Time_Group)) +
  geom_bar(color = "#238B45", fill = "#238B45") +
  theme_bw() +
  facet_wrap( ~ Class, scales = "free", ncol = 2)
```

This will generate the following output:

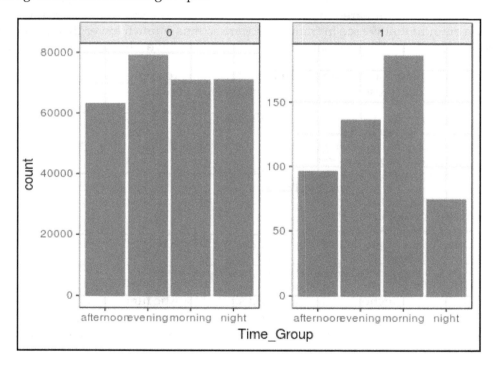

The inference obtained from this visualization is that the number of non-fraudulent transactions remains almost the same across all time periods of the day, whereas we see a huge rise in the number of fraudulent transactions during the morning Time group.

Let's do a last bit of exploration of the transaction amount with respect to class:

```
# getting the summary of amount with respect to the class
print(tapply(cc_fraud$Amount  ,cc_fraud$Class,summary))
```

The preceding code will generate the following output:

```
$`0`
   Min.  1st Qu.   Median    Mean 3rd Qu.      Max.
   0.00     5.65    22.00   88.29   77.05 25691.16
$`1`
   Min. 1st Qu.  Median   Mean 3rd Qu.     Max.
   0.00    1.00    9.25 122.21 105.89 2125.87
```

One interesting insight from the summary is that the mean amount in fraudulent transactions is higher compared to genuine transactions. However, the maximum transaction amount that we see in fraudulent transactions is much lower than the genuine transactions. It can also be seen that genuine transactions have a higher median amount.

Now, let's convert our R dataframe to an H2O dataframe to apply the AE to it. This is a requirement in order to use the functions from the h2o library:

```
# converting R dataframe to H2O dataframe
cc_fraud_h2o <- as.h2o(cc_fraud)
#splitting the data into 60%, 20%, 20% chunks to use them as training,
#vaidation and test datasets
splits <- h2o.splitFrame(cc_fraud_h2o,ratios = c(0.6, 0.2), seed = 148)
# creating new train, validation and test h2o dataframes
train <- splits[[1]]
validation <- splits[[2]]
test <- splits[[3]]
# getting the target and features name in vectors
target <- "Class"
features <- setdiff(colnames(train), target)
```

The tanh activation function is a rescaled and shifted logistic function. Other functions, such as ReLu and Maxout, are also provided by the h2o library and they can also be used. In the first AE model, let's use the tanh activation function. This choice is arbitrary and other activation functions may also be tried as desired.

The h2o.deeplearning function has a parameter AE and this should be set to TRUE to train a AE model. Let's build our AE model now:

```
model_one = h2o.deeplearning(x = features, training_frame = train,
                             AE = TRUE,
                             reproducible = TRUE,
                             seed = 148,
```

```
                                    hidden = c(10,10,10), epochs = 100,
activation = "Tanh",
                                    validation_frame = test)
```

The preceding code generates the following output:

```
|==========================================================================
================================================| 100%
```

We will save the model so we do not have to retrain t again and again. Then load the model
that is persisted on the disk and print the model to verify the AE learning using the
following code:

```
h2o.saveModel(model_one, path="model_one", force = TRUE)
model_one<-
h2o.loadModel("/home/sunil/model_one/DeepLearning_model_R_1544970545051_1")
print(model_one)
```

This will generate the following output:

```
Model Details:
==============
H2OAutoEncoderModel: deeplearning
Model ID:  DeepLearning_model_R_1544970545051_1
Status of Neuron Layers: auto-encoder, gaussian distribution, Quadratic
loss, 944 weights/biases, 20.1 KB, 2,739,472 training samples, mini-batch
size 1
   layer units  type dropout       l1        l2 mean_rate rate_rms momentum
mean_weight weight_rms mean_bias bias_rms
1     1   34 Input  0.00 %        NA        NA        NA       NA       NA
NA        NA        NA        NA
2     2   10  Tanh  0.00 % 0.000000 0.000000  0.610547 0.305915 0.000000
-0.000347   0.309377 -0.028166 0.148318
3     3   10  Tanh  0.00 % 0.000000 0.000000  0.181705 0.103598 0.000000
0.022774   0.262611 -0.056455 0.099918
4     4   10  Tanh  0.00 % 0.000000 0.000000  0.133090 0.079663 0.000000
0.000808   0.337259  0.032588 0.101952
5     5   34  Tanh     NA 0.000000 0.000000  0.116252 0.129859 0.000000
0.006941   0.357547  0.167973 0.688510
H2OAutoEncoderMetrics: deeplearning
 Reported on training data.
 Training Set Metrics:
 =====================
MSE: (Extract with `h2o.mse`) 0.0003654009
RMSE: (Extract with `h2o.rmse`) 0.01911546
H2OAutoEncoderMetrics: deeplearning
 Reported on validation data.
 Validation Set Metrics:
```

```
=====================
MSE: (Extract with `h2o.mse`) 0.0003508435
RMSE: (Extract with `h2o.rmse`) 0.01873082
```

We will now make predictions on test dataset using the AE model that is built, using the following code:

```
test_autoencoder <- h2o.predict(model_one, test)
```

This will generate the following output:

```
|=====================================================================
=========================================| 100%
```

It is possible to visualize the encoder representing the data in a conscious manner in the inner layers through the `h2o.deepfeatures` function. Let's try visualizing the reduced data in a second layer:

```
train_features <- h2o.deepfeatures(model_one, train, layer = 2) %>%
  as.data.frame() %>%
  mutate(Class = as.vector(train[, 31]))
# printing the reduced data represented in layer2
print(train_features%>%head(3))
```

The preceding code will generate the following output:

```
DF.L2.C1    DF.L2.C2      DF.L2.C3      DF.L2.C4     DF.L2.C5
-0.12899115 0.1312075   0.115971952 -0.12997648 0.23081912
-0.10437942 0.1832959   0.006427409 -0.08018725 0.05575977
-0.07135827 0.1705700  -0.023808057 -0.11383244 0.10800857
DF.L2.C6    DF.L2.C7     DF.L2.C8  DF.L2.C9  DF.L2.C10  Class0.1791547
0.10325721   0.05589069 0.5607497 -0.9038150      0
0.1588236 0.11009450 -0.04071038 0.5895413 -0.8949729        0
0.1676358 0.10703990 -0.03263755 0.5762191 -0.8989759        0
```

Let us now plot the data of `DF.L2.C1` with respect to `DF.L2.C2` to verify if the encoder has detected the fraudulent transactions, using the following code:

```
ggplot(train_features, aes(x = DF.L2.C1, y = DF.L2.C2, color = Class)) +
  geom_point(alpha = 0.1,size=1.5)+theme_bw()+
  scale_fill_brewer(palette = "Accent")
```

This will generate the following output:

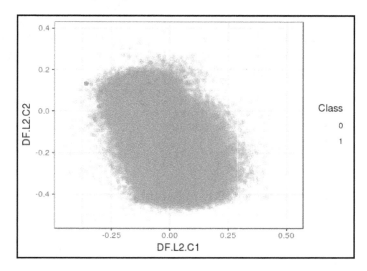

Again we plot the data of DF.L2.C3 with respect to DF.L2.C4 to verify the if the encoder have detected any fraud transaction, using the following code:

```
ggplot(train_features, aes(x = DF.L2.C3, y = DF.L2.C4, color = Class)) +
  geom_point(alpha = 0.1,size=1.5)+theme_bw()+
  scale_fill_brewer(palette = "Accent")
```

The preceding code will generate the following output:

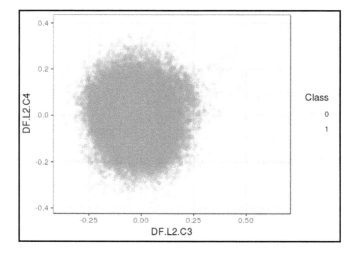

We see from the two visualizations that the fraudulent transactions are indeed detected by the dimensionality reduction approach with our AE model. Those few scattered dots (represented by 1) depicts the fraud transactions that are detected. We can also train a new model with the other hidden layers, using our first model. This results in 10 columns, since the third layer has 10 nodes. We are just attempting to slice out one layer where some level of reduction was done and use that to build a new model:

```
# let's consider the third hidden layer. This is again a random choice
# in fact we could have taken any layer among the 10 inner layers
train_features <- h2o.deepfeatures(model_one, validation, layer = 3) %>%
  as.data.frame() %>%
  mutate(Class = as.factor(as.vector(validation[, 31]))) %>%
  as.h2o()
```

The preceding code will generate the following output:

```
|========================================================================
=================================================| 100%
|========================================================================
=================================================| 100%
```

As we can see, the training models and data are successfully created. We will now go ahead and train the new model, save it and the print it. First, we will get the feature names from the sliced encoder layer:

```
features_two <- setdiff(colnames(train_features), target)
```

Then we will training a new model:

```
model_two <- h2o.deeplearning(y = target,
                              x = features_two,
                              training_frame = train_features,
                              reproducible = TRUE,
                              balance_classes = TRUE,
                              ignore_const_cols = FALSE,
                              seed = 148,
                              hidden = c(10, 5, 10),
                              epochs = 100,
                              activation = "Tanh")
```

We will then save the model to avoid retraining again, then retrieve the model and print it using the following code:

```
h2o.saveModel(model_two, path="model_two", force = TRUE)
model_two <-
h2o.loadModel("/home/sunil/model_two/DeepLearning_model_R_1544970545051_2")
print(model_two)
```

This will generate the following output:

```
Model Details:
==============
H2OBinomialModel: deeplearning
Model ID:  DeepLearning_model_R_1544970545051_2
Status of Neuron Layers: predicting Class, 2-class classification,
bernoulli distribution, CrossEntropy loss, 247 weights/biases, 8.0 KB,
2,383,962 training samples, mini-batch size 1
   layer units     type dropout        l1        l2 mean_rate rate_rms momentum
mean_weight weight_rms mean_bias bias_rms
1    1    10    Input  0.00 %        NA        NA        NA        NA        NA
NA        NA        NA        NA
2    2    10     Tanh  0.00 % 0.000000 0.000000  0.001515 0.001883 0.000000
-0.149216   0.768610 -0.038682 0.891455
3    3     5     Tanh  0.00 % 0.000000 0.000000  0.003293 0.004916 0.000000
-0.251950   0.885017 -0.307971 0.531144
4    4    10     Tanh  0.00 % 0.000000 0.000000  0.002252 0.001780 0.000000
0.073398   1.217405 -0.354956 0.887678
5    5     2 Softmax        NA 0.000000 0.000000  0.007459 0.007915 0.000000
-0.095975   3.579932  0.223286 1.172508
H2OBinomialMetrics: deeplearning
 Reported on training data.
  Metrics reported on temporary training frame with 9892 samples
 MSE:   0.1129424
RMSE:   0.336069
LogLoss:  0.336795
Mean Per-Class Error:  0.006234916
AUC:  0.9983688
Gini:  0.9967377
Confusion Matrix (vertical: actual; across: predicted) for F1-optimal
threshold:
        0    1    Error      Rate
0    4910   62 0.012470   =62/4972
1       0 4920 0.000000   =0/4920
Totals 4910 4982 0.006268   =62/9892
Maximum Metrics: Maximum metrics at their respective thresholds
                      metric threshold    value idx
1                     max f1  0.009908 0.993739 153
2                     max f2  0.009908 0.997486 153
3                max f0point5  0.019214 0.990107 142
4                max accuracy  0.009908 0.993732 153
5               max precision  1.000000 1.000000   0
6                  max recall  0.009908 1.000000 153
7             max specificity  1.000000 1.000000   0
8            max absolute_mcc  0.009908 0.987543 153
9   max min_per_class_accuracy  0.019214 0.989541 142
10 max mean_per_class_accuracy  0.009908 0.993765 153
```

```
Gains/Lift Table: Extract with `h2o.gainsLift(<model>, <data>)` or
`h2o.gainsLift(<model>, valid=<T/F>, xval=<T/F>)
```

For measuring model performance on test data, we need to convert the test data to the same reduced dimensions as the training data:

```
test_3 <- h2o.deepfeatures(model_one, test, layer = 3)
print(test_3%>%head())
```

The preceding code will generate the following output:

```
| =========================================================================
=================================================| 100%
```

We see, the data has been converted successfully. Now, to make predictions on the test dataset with `model_two`, we will use the following code:

```
test_pred=h2o.predict(model_two, test_3,type="response")%>%
    as.data.frame() %>%
    mutate(actual = as.vector(test[, 31]))
```

This will generate the following output:

```
| =========================================================================
=================================================| 100%
```

As we can see, from the output, predictions has been successfully completed and now let us visualize the predictions using the following code:

```
test_pred%>%head()
  predict          p0          p1 actual
1       0 1.0000000 1.468655e-23      0
2       0 1.0000000 2.354664e-23      0
3       0 1.0000000 5.987218e-09      0
4       0 1.0000000 2.888583e-23      0
5       0 0.9999988 1.226122e-06      0
6       0 1.0000000 2.927614e-23      0
# summarizing the predictions
print(h2o.predict(model_two, test_3) %>%
    as.data.frame() %>%
    dplyr::mutate(actual = as.vector(test[, 31])) %>%
    group_by(actual, predict) %>%
    dplyr::summarise(n = n()) %>%
    mutate(freq = n / sum(n)))
```

This will generate the following output:

```
|================================================================
================================================|  100%
# A tibble: 4 x 4
# Groups:    actual [2]
  actual predict     n    freq
  <chr>  <fct>    <int>   <dbl>
1 0      0        55811  0.986
2 0      1          817  0.0144
3 1      0           41  0.414
4 1      1           58  0.586
```

We see that our AE is able to correctly predict non-fraudulent transactions with 98% accuracy, which is good. However, it is yielding only 58% accuracy when predicting fraudulent transactions. This is definitely something to focus on. Our model needs some improvement, and this can be accomplished through the following options:

- Using other layers' latent space representations as input to build `model_two` (recollect that we currently use the layer 3 representation)
- Using ReLu or Maxout activation functions instead of `Tanh`
- Checking the misclassified instances through the `h2o.anomaly` function and increasing or decreasing the cutoff threshold MSE values, which separates the fraudulent transactions from non-fraudulent transactions
- Trying out a more complex architecture in the encoder and decoder

We are not going to be attempting these options in this chapter as they are experimental in nature. However, interested readers may try and improve the accuracy of the model by trying these options.

Finally, one best practice is to explicitly shut down the `h2o` cluster. This can be accomplished with the following command:

```
h2o.shutdown()
```

Summary

In this chapter, we learned about an unsupervised deep learning technique called AEs. We covered the definition, working principle, types, and applications of AEs. H2O, an open source library that enables us to create deep learning models, including AEs, was explored. We then discussed a credit card fraud open dataset and implemented a project with an AE to detect fraudulent credit card transactions.

Can deep neural networks help with creative tasks such as prose generation, story writing, caption generation for images, and poem writing? Not sure?! Let's explore RNNs, in the next chapter, a special type of deep neural network that enables us to accomplish creative tasks. Turn the page to explore the world of RNNs for prose generation.

8

Automatic Prose Generation with Recurrent Neural Networks

We have been interacting through this book for almost 200 pages, but I realized that I have not introduced myself properly to you! I guess it's time. You already know some bits about me through the author profile in this book; however, I want to tell you a bit about the city I live in. I am based in South India, in a city called Bengaluru, also know as Bangalore. The city is known for its IT talent and population diversity. I love the city, as it is filled with loads of positive energy. Each day, I get to meet people from all walks of life—people from multiple ethnicities, multiple backgrounds, people who speak multiple languages, and so on. Kannada is the official language spoken in the state of Karnataka where Bangalore is located. Though I can speak bits and pieces of Kannada, my proficiency with speaking the language is not as good as a native Kannada speaker. Of course, this is an area of improvement for me and I am working on it. Like me, many other migrants that moved to the city from other places also face problems while conversing in Kannada. Interestingly, not knowing the language does not stop any of us from interacting with locals in their own language. Guess what comes to our rescue: mobile apps such as Google translate, Google text-to-speech, and the like. These applications are built on NLP technologies called machine translation and speech recognition. These technologies in turn work on things known as **language models**. Language models is the topic we will delve into in this chapter.

The objectives of the chapter include exploring the following topics:

- The need for language modeling to address natural language processing tasks
- The working principle of language models
- Application of language models
- Relationship between language modeling and neural networks
- Recurrent neural networks

- Differences between a normal feedforward network and a recurrent neural network
- Long short-term memory networks
- A project to autogenerate text using recurrent neural networks

Understanding language models

In the English language, the character *a* appears much more often in words and sentences than the character *x*. Similarly, we can also observe that the word *is* occurs more frequently than the word *specimen*. It is possible to learn the probability distributions of characters and words by examining large volumes of text. The following screenshot is a chart showing the probability distribution of letters given a corpus (text dataset):

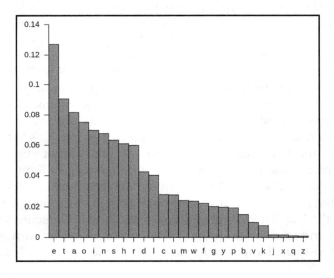

Probability distribution of letters in a corpus

We can observe that the probability distributions of characters are non-uniform. This essentially means that we can recover the characters in a word, even if they are lost due to noise. If a particular character is missing in a word, it can be reconstructed just based on the characters that are surrounding the missing character. The reconstruction of the missing character is not done randomly, but is done by picking the character that has the highest probability distribution of occurrence, given the characters that are surrounding the missing character. Technically speaking, the statistical structure of words in a sentence or characters in words follows the distance from maximal entropy.

A language model exploits the statistical structure of a language to express the following:

- Given w_1, w_2, w_3,...w_N words in a sentence, a language model assigns a probability to a sentence P(w_1, w_2, w_3,.... w_N).
- It then assigns probability of an upcoming word (w_4 in this case) as P(w_4 | w_1, w_2, w_3).

Language models enable a number of applications to be developed in NLP, and some of them are listed as follows:

- **Machine translation**: P(enormous cyclone tonight) > P(gain typhoon this evening)
- **Spelling correction**: P(satellite constellation) > P(satelitte constellation)
- **Speech recognition**: P(I saw a van) > P(eyes awe of an)
- **Typing prediction**: Auto completion of in Google search, typing assistance apps

Let's now look at how the probabilities are calculated for the words. Consider a simple sentence, *Decembers are cold*. The probability of this sentence is expressed as follows:

*P("Decembers are cold") = P("December") * P ("are" | "Decembers") * P("cold" | "Decembers are")*

Mathematically, the probability computation of words in a sentence (or letters in a word) can be expressed as follows:

$$P(w_1, w_2, w_3, w_4, w_5, \cdots, w_n) = \prod_{i=1}^{n} P(w_1 | w_1, w_2, w_3, w_4, w_5, \cdots, w_{i-1})$$

Andrey Markov, a Russian mathematician, described a stochastic process with a property called **Markov Property** or **Markov Assumption**. This basically states that one can make predictions for the future of the process based solely on its present state, just as well as one could knowing the process's full history, hence independently from such history.

Based on Markov's assumption, we can rewrite the conditional probability of *cold* as follows:

P("cold" | "Decembers are") is congruent to P("cold" | "are")

Mathematically, Markov's assumption can be expressed as follows:

$$P(w_n | w_1, w_2, w_3, w_4, w_5, \cdots, w_{n-1}) = P(w_n | w_{n-1})$$

While this mathematical formulation represents the bigram model (two words taken into consideration at a time), it can be easily extended to an n-gram model. In the n-gram model, the conditional probability depends on just a couple more previous words.

Mathematically, an n-gram model is expressed as follows:

$$P(w_n | w_1, w_2, w_3, w_4, w_5, \cdots, w_{n-1}) = P(w_n | w_{n-k}, \cdots, w_{n-1})$$

Consider the famous poem *A Girl* by *Ezra Pound* as our corpus for building a **bigram** model. The following is the text corpus:

```
The tree has entered my hands,
The sap has ascended my arms,
The tree has grown in my breast-Downward,
The branches grow out of me, like arms.
Tree you are,
Moss you are,
You are violets with wind above them.
A child - so high - you are,
And all this is folly to the world.
```

We are already aware that in a bigram model, the conditional probability is computed just based on the previous word. So, the probability of a word can be computed as follows:

$$P(w_i | w_{i-1}) = \frac{count(w_{i-1}, w_i)}{count(w_{i-1})}$$

If we were to compute the probability of the word *arms* given the word *my* in the poem, it is computed as the number of times the words *arms* and *my* appear together in the poem, divided by the number of times the word *my* appears in the poem.

We see that the words *my arms* appeared in the poem only once (in the sentence *The sap has ascended my arms*). However, the word *my* appeared in the poem three times (in the sentences *The tree has entered my hands, The sap has ascended my arms,* and *The tree has grown in my breast-Downward*).

Therefore, the conditional probability of the word *arms* given *my* is 1/3, formally represented as follows:

P("arms" | "my") = P("arms", "my") / P("my") = 1 / 3

To calculate probability of the first and last words, the special tags <BOS> and <EOS> are added at the start and end of sentences, respectively. Similarly, the probability of a sentence or sequence of words can be calculated using the same approach by multiplying all the bigram probabilities.

As language modeling involves predicting the next word in a sequence, given the sequence of words already present, we can train a language model to create subsequent words in a sequence from a given starting sequence.

Exploring recurrent neural networks

Recurrent neural networks (RNNs) are a family of neural networks for processing sequential data. RNNs are generally used to implement language models. We, as humans, base much of our language understanding on the context. For example, let's consider the sentence *Christmas falls in the month of* --------. It is easy to fill in the blank with the word *December*. The essential idea here is that there is information about the last word encoded in the previous elements of the sentence.

The central theme behind the RNN architecture is to exploit the sequential structure of the data. As the name suggests, RNNs operate in a recurrent way. Essentially, this means that the same operation is performed for every element of a sequence or sentence, with its output depending on the current input and the previous operations.

An RNN works by looping an output of the network at time *t* with the input of the network at time *t+1*. These loops allow persistence of information from one time step to the next one. The following diagram is a circuit diagram representing an RNN:

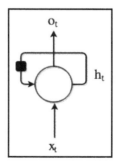

Circuit diagram representing a RNN

The diagram indicates an RNN that remembers what it knows from previous input using a simple loop. This loop takes the information from the previous timestamp and adds it to the input of the current timestamp. At a particular time step t, X_t is the input to the network, O_t is the output of the network, and h_t is the detail it remembered from previous nodes in the network. In between, there is the RNN cell, which contains neural networks just like a feedforward network.

 One key point to ponder in terms of the definition of an RNN is the timestamps. The timestamps referred to in the definition have nothing to do with past, present, and future. They simply represent a word or an item in a sequence or a sentence.

Let's consider an example sentence: *Christmas Holidays are Awesome*. In this sentence, take a look at the following timestamp:

- *Christmas* is x_0
- *Holidays* is x_1
- *are* is x_2;
- *Awesome* is x_3

If t=1, then take a look at the following:

- x_t = *Holidays* → event at current timestamp
- x_{t-1} = *Christmas* → event at previous timestamp

It can be observed from the preceding circuit diagram that the same operation is performed in the RNN repeatedly on different nodes. There is also a black square in the diagram that represents a time delay of a single time step. It may be confusing to understand the RNN with the loops, so let's unfold the computational graph. The unfolded RNN computational graph is shown in the following diagram:

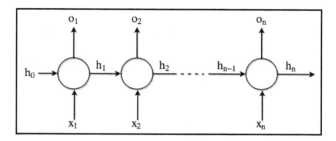

RNN—unfolded computational graph view

In the preceding diagram, each node is associated with a particular time. In the RNN architecture, each node receives different inputs at each time step x_t. It also has the capability of producing outputs at each time step o_t. The network also maintains a memory state h_t, which contains information about what happened in the network up to the time t. As this is the same process that is run across all the nodes in the network, it is possible to represent the whole network in a simplified form, as shown in the RNN circuit diagram.

Now, we understand that we see the word **recurrent** in RNNs because it performs the same task for every element of a sequence, with the output depending on previous computations. It may be noted that, theoretically, RNNs can make use of information in arbitrarily long sequences, but in practice, they are implemented to looking back only a few steps.

Formally, an RNN can be defined in an equation as follows:

$$h_t = \phi(WX_t + Uh_{t-1})$$

In the equation, h_t is the hidden state at timestamp t. An activation function such as Tanh, Sigmoid, or ReLU can be applied to compute the hidden state and it is represented in the equation as ϕ. W is the weight matrix for the input to the hidden layer at timestamp t. X_t is the input at timestamp t. U is the weight matrix for the hidden layer at time t-1 to the hidden layer at time t, and h_{t-1} is the hidden state at timestamp t.

During backpropagation, U and W weights are learned by the RNN. At each node, the contribution of the hidden state and the contribution of the current input are decided by U and W. The proportions of U and W in turn result in the generation of output at the current node. The activation functions add non-linearity in RNNs, thus enabling the simplification of gradient calculations during the backpropagation process. The following diagram illustrates the idea of backpropagation:

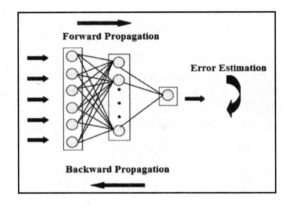

Backpropagation in neural networks

The following diagram depicts the overall working mechanism of an RNN and the way the weights U and W are learned through backpropagation. It also depicts the use of the U and W weight matrices in the network to generate the output, as shown in the following diagram:

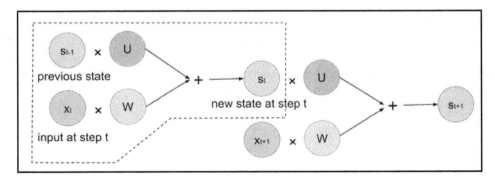

Role of weights in RNN

Comparison of feedforward neural networks and RNNs

One fundamental difference between other neural networks and RNNs is that, in all other networks, the inputs are independent of each other. However, in an RNN, all the inputs are related to each other. In an application, to predict the next word in a given sentence, the relationship between all the previous words helps to predict the current output. In other words, an RNN remembers all these relationships while training itself. This is not the case with other types of neural networks. A representation of a feedforward network is illustrated in the following diagram:

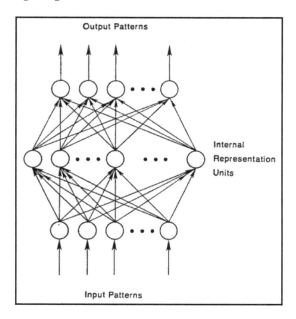

Feedforward neural network architecture

From the preceding diagram, we can see that no loops are involved in the feedforward network architecture. This is in contrast to that of the RNN architecture depicted in the diagrams of the RNN circuit diagram and the RNN unfolded computational graph. The series of mathematical operations in a feedforward network is performed at the nodes and the information is processed straight through, with no loops whatsoever.

Using supervised learning, the input that is fed to a feedforward network is transformed into output. The output in this case could be a label if it is classification, or it is a number in the case of regression. If we consider image classification, a label can be *cat* or *dog* for an image given as input.

A feedforward neural network is trained on labeled images until errors are minimized in predicting the labels. Once trained, the model is able to classify even images that it has not seen previously. A trained feedforward network can be exposed to any random collection of photographs; the categorization of the first photograph does not have any impact or influence on the second or subsequent photographs that the model needs to categorize. Let's discuss this with an example for better clarity on the concept: if the first image is seen as a *dog* by the feedforward network, it does not imply that the second image will be classified as *cat*. In other words, the predictions that the model arrives at have no notion of order in time, and the decision regarding the label is arrived at just based on the current input that is provided. To summarize, in feedforward networks no information on historical predictions is used for current predictions. This is very different from RNNs, where the previous prediction is considered in order to aid the current prediction.

Another important difference is that feedforward networks, by design, map one input to one output, whereas RNNs can take multiple forms: map one input to multiple outputs, many inputs to many outputs, or many inputs to one output. The following diagram depicts the various input-output mappings possible with RNNs:

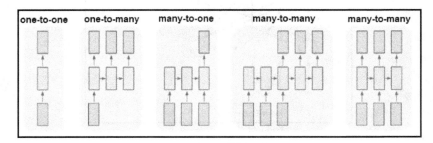

Input-output mapping possibilities with RNNs

Let's review some of the practical applications of the various input-output mappings possible with an RNN. Each rectangle in the preceding diagram is a vector and the arrows represent functions, for example, a matrix multiplication. The input vectors are the lower rectangles (colored in red), and the output vectors are the upper rectangles (colored in blue color). The middle rectangles (colored in green) are vectors that hold the RNN's state.

The following are the various forms of mapping illustrated in the diagram:

- **One input to one output**: The leftmost one is a vanilla mode of processing without RNN, from fixed-sized input to fixed-sized output; for example, image classification.
- **One input to many outputs**: Sequence output, for example, image captioning takes an image as input and it outputs a sentence of words.
- **Many inputs to one output**: Sequence input, for example, sentiment analysis where a given sentence is given as input to the RNN, and the output is a classification expressing positive or negative sentiment of the sentence.
- **Many inputs to many outputs**: Sequence input and sequence output; for example, for a machine translation task, an RNN reads a sentence in English as input and then outputs a sentence in Hindi or some other language.
- **Many inputs to many outputs**: Synced sequence input and output, for example, video classification where we wish to label each frame of the video.

Let's now review the final difference between a feedforward network and an RNN. The way backpropagation is done in order to set the weights in a feedforward neural network is different from that of what is called **backpropagation through time** (**BPTT**), which is carried out in an RNN. We are already aware that the objective of the backpropagation algorithm in neural networks is to adjust the weights of a neural network to minimize the error of the network outputs compared to some expected output in response to corresponding inputs. Backpropagation itself is a supervised learning algorithm that allows the neural network to be corrected with regard to the specific errors made. The backpropagation algorithm involves the following steps:

1. Provide training input to the neural network and propagate it through the network to get the output
2. Compare the predicted output to the actual output and calculate the error
3. Calculate the derivatives of the error with respect to the learned network weights
4. Modify the weights to minimize the error
5. Repeat

In feedforward networks, it makes sense to run backpropagation at the end as the output is available only at the end. In RNNs, the output is produced at each time step and this output influences the output in the subsequent time steps. In other words, in RNNs, the error of a time step depends on the previous time step. Therefore, the normal backpropagation algorithms are not suitable for RNNs. Hence, a different algorithm known as BPTT is used to modify the weights in an RNN.

Backpropagation through time

We are already aware that RNNs are cyclical graphs, unlike feedforward networks, which are acyclic directional graphs. In feedforward networks, the error derivatives are calculated from the layer above. However, in an RNN we don't have such layering to perform error derivative calculations. A simple solution to this problem is to unroll the RNN and make it similar to a feedforward network. To enable this, the hidden units from the RNN are replicated at each time step. Each time step replication forms a layer that is similar to layers in a feedforward network. Each time step t layer connects to all possible layers in the time step $t+1$. Therefore, we randomly initialize the weights, unroll the network, and then use backpropagation to optimize the weights in the hidden layer. The lowest layer is initialized by passing parameters. These parameters are also optimized as a part of backpropagation. The backpropagation through time algorithm involves the following steps:

1. Provide a sequence of time steps of input and output pairs to the network
2. Unroll the network, then calculate and accumulate errors across each time step
3. Roll up the network and update weights
4. Repeat

In summary, with BPTT, the error is back propagated from the last to the first time step, while unrolling all the time steps. The error for each time step is calculated, which allows updating the weights. The following diagram is a visualization showing the backpropagation through time:

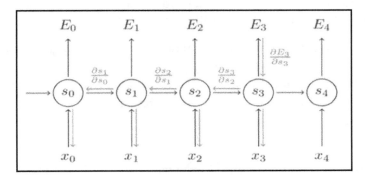

Backpropagation through time in an RNN

It should be noted that as the number of time steps increases, the BPTT algorithm can get computationally very expensive.

Problems and solutions to gradients in RNN

RNNs are not perfect, there are two main issues namely **exploding gradients** and **vanishing gradients** that they suffer from. To understand the issues, let's first understand what a gradient means. A gradient is a partial derivative with respect to its inputs. In simple layman's terms, a gradient measures how much the output of a function changes, if one were to change the inputs a little bit.

Exploding gradients

Exploding gradients relate to a situation where the BPTT algorithm assigns an insanely high importance to the weights, without a rationale. The problem results in an unstable network. In extreme situations, the values of weights can become so large that the values overflow and result in NaN values.

The exploding gradients problem can be detected through observing the following subtle signs while training the network:

- The model weights quickly become very large during training
- The model weights become NaN values during training
- The error gradient values are consistently above 1.0 for each node and layer during training

There are several ways in which one could handle the exploding gradients problem. The following are some of the popular techniques:

- This problem can be easily solved if we can truncate or squash the gradients. This is known as **gradient clipping**.
- Updating weights across fewer prior time steps during training may also reduce the exploding gradient problem. This technique of having fewer step updates is called **truncated backpropagation through time** (**TBPTT**). It is an altered version of the BPTT training algorithm where the sequence is processed one time step at a time, and periodically ($k1$ time steps) the BPTT update is performed back for a fixed number of time steps ($k2$ time steps). $k1$ is the number of forward-pass time steps between updates. $k2$ is the number of time steps to which to apply BPTT.
- Weight regularization can be done by checking the size of network weights and applying a penalty to the networks loss function for large weight values.

- By using **long short-term memory units (LSTMs)** or **gated recurrent units (GRUs)** instead of plain vanilla RNNs.
- Careful initialization of weights such as **Xavier** initialization or **He** initialization.

Vanishing gradients

We are already aware that long-term dependencies are very important for RNNs to function correctly. RNNs can become too deep because of the long-term dependencies. The vanishing gradient problem arises in cases where the gradient of the activation function is very small. During backpropagation, when the weights are multiplied with the low gradients, they tend to become very small and vanish as they go further into the network. This makes the neural network forget the long-term dependency. The following diagram is an illustration showing the cause that leads to vanishing gradients:

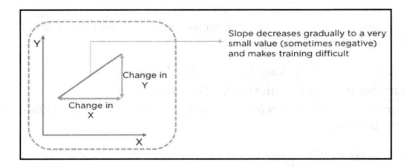

The cause of vanishing gradients

To summarize, due to the vanishing gradients problem, RNNs experience difficulty in memorizing previous words very far away in the sequence and are only able to make predictions based on the most recent words. This can impact the accuracy of RNN predictions. At times, the model may fail to predict or classify what it is supposed to do.

There are several ways in which one could handle the vanishing gradients problem. The following are some of the most popular techniques:

- Initialize network weights for the identity matrix so that the potential for a vanishing gradient is minimized.

- Setting the activation functions to ReLU instead of `sigmoid` or `tanh`. This makes the network computations stay close to the identity function. This works well because when the error derivatives are being propagated backwards through time, they remain constants of either 0 or 1, and so aren't likely to suffer from vanishing gradients.
- Using LSTMs, which are a variant of the regular recurrent network designed to make it easy to capture long-term dependencies in sequence data. The standard RNN operates in such a way that the hidden state activation is influenced by the other local activations closest to it, which corresponds to a **short-term memory**, while the network weights are influenced by the computations that take place over entire long sequences, which corresponds to a **long-term memory**. The RNN was redesigned so that it has an activation state that can also act like weights and preserve information over long distances, hence the name **long short-term memory**.

In LSTMs, rather than each hidden node being simply a node with a single activation function, each node is a memory cell in itself that can store other information. Specifically, it maintains its own cell state. Normal RNNs take in their previous hidden state and the current input, and output a new hidden state. An LSTM does the same, except it also takes in its old cell state and will output its new cell state.

Building an automated prose generator with an RNN

In this project, we will attempt to build a character-level language model using an RNN to generate prose given some initial seed characters. The main task of a character-level language model is to predict the next character given all previous characters in a sequence of data. In other words, the function of an RNN is to generate text character by character.

To start with, we feed the RNN a huge chunk of text as input and ask it to model the probability distribution of the next character in the sequence, given a sequence of previous characters. These probability distributions conceived by the RNN model will then allow us to generate new text, one character at a time.

The first requirement for building a language model is to secure a corpus of text that the model can use to compute the probability distribution of various characters. The larger the input text corpus, the better the RNN will model the probabilities.

We do not have to strive a lot to secure the big text corpus that is required to train the RNN. There are classical texts (books) such as *The Bible* that can be used as a corpus. The best part is many of the classical texts are no longer protected under copyright. Therefore, the texts can be downloaded and used freely in our models.

Project Gutenberg is the best place to get access to free books that are no longer protected by copyright. Project Gutenberg can be accessed through the URL `http://www.gutenberg.org`. There are several books such as *The Bible, Alice's Adventures in Wonderland*, and so on are available from Project Gutenberg. As of December 2018, there are 58,486 books available for download. The books are available in several formats for us to be able to download and use, not just for this project, but for any project where huge text corpus input is required. The following screenshot is of a sample book from Project Gutenberg and the various formats in which the book is available for download:

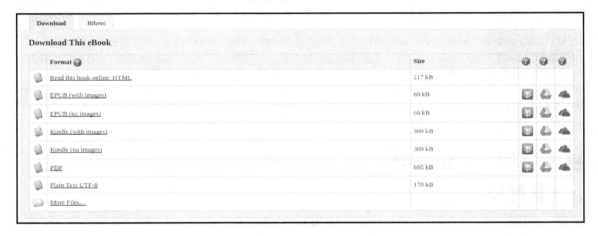

Sample book available from Project Gutenberg in various formats

Irrespective of the format of the file that is downloaded, Project Gutenberg adds standard header and footer text to the actual book text. The following is an example of the header and footer that can be seen in a book:

```
*** START OF THIS PROJECT GUTENBERG EBOOK ALICE'S ADVENTURES IN WONDERLAND ***

THE END
```

It is essential that we remove this header and footer text from the book text downloaded from Project Gutenberg website. For a text file that is downloaded, one can open the file in a text editor and delete the header and footer.

For our project in this chapter, let's use a favorite book from childhood as the text corpus: *Alice's Adventures in Wonderland* by Lewis Carroll. While we have an option to download the text format of this book from Project Gutenberg and make use of it as a text corpus, the R language's languageR library makes the task even more easier for us. The languageR library already has the *Alice's Adventures in Wonderland* book text. After installing the languageR library, use the following code to load the text data into the memory and print the loaded text:

```
# including the languageR library
library("languageR")
# loading the "Alice's Adventures in Wonderland" to memory
data(alice)
# printing the loaded text
print(alice)
```

You will get the following output:

```
[1] "ALICE"         "S"          "ADVENTURES"     "IN"
"WONDERLAND"
[6] "Lewis"         "Carroll"      "THE"           "MILLENNIUM"
"FULCRUM"
   [11] "EDITION"    "3"           "0"             "CHAPTER"
"I"
   [16] "Down"       "the"         "Rabbit-Hole"    "Alice"
"was"
   [21] "beginning"  "to"          "get"           "very"
"tired"
   [26] "of"         "sitting"     "by"            "her"
"sister"
   [31] "on"         "the"         "bank"          "and"
"of"
   [36] "having"     "nothing"     "to"            "do"
"once"
   [41] "or"         "twice"       "she"           "had"
"peeped"
   [46] "into"       "the"         "book"          "her"
"sister"
   [51] "was"        "reading"     "but"           "it"
"had"
   [56] "no"         "pictures"    "or"            "conversations"
"in"
```

We see from the output that the book text is stored as a character vector, where each of the vector items is a word from the book text that is split by punctuation. It may also be noted that not all the punctuation is retained in the book text.

The following code reconstructs the sentences from the words in the character vector. Of course, we do not get things like sentence boundaries during the reconstruction process, as the character vector does not have as much punctuation as character vector items. Now, let's do the reconstruction of the book text from individual words:

```
alice_in_wonderland<-paste(alice,collapse=" ")
print(alice_in_wonderland)
```

You will get the following output:

```
[1] "ALICE S ADVENTURES IN WONDERLAND Lewis Carroll THE MILLENNIUM FULCRUM
EDITION 3 0 CHAPTER I Down the Rabbit-Hole Alice was beginning to get very
tired of sitting by her sister on the bank and of having nothing to do once
or twice she had peeped into the book her sister was reading but it had no
pictures or conversations in it and what is the use of a book thought Alice
without pictures or conversation So she was considering in her own mind as
well as she could for the hot day made her feel very sleepy and stupid
whether the pleasure of making a daisy-chain would be worth the trouble of
getting up and picking the daisies when suddenly a White Rabbit with pink
eyes ran close by her There was nothing so VERY remarkable in that nor did
Alice think it so VERY much out of the way to hear the Rabbit say to itself
Oh dear Oh dear I shall be late when she thought it over afterwards it
occurred to her that she ought to have wondered at this but at the time it
all seemed quite natural but when the Rabbit actually TOOK A WATCH OUT OF
ITS WAISTCOAT- POCKET and looked at it and then hurried on Alice started to
her feet for it flashed across her mind that she had never before seen a
rabbit with either a waistcoat-pocket or a watch to take out of it and
burning with curiosity she ran across the field after it and fortunately
was just in time to see it pop down a large rabbit-hole under the hedge In
another moment down went Alice after it never once considering how in the
world she was to get out again The rabbit-hole we .......
```

From the output, we see that a long string of text is constructed from the words. Now, we can move on to doing some preprocessing on this text to feed it to the RNN so that the model learns the dependencies between characters and the conditional probabilities of characters in sequences.

One of the things to note is that, as with a character-level language model that generates the next character in a sequence, you can build a word-level language model too. However, the character-level language model has an advantage in that it can create its own unique words that are not in the vocabulary we train it on.

Let's now learn how RNN works to conceive the dependencies between characters in sequences. Assume that we only had a vocabulary of four possible letters, [a, p, l, e], and the intent is to train an RNN on the training sequence *apple*. This training sequence is in fact a source of four separate training examples:

- The probability of the letter *p* should be likely, given the context of *a, ,* in other words, the conditional probability of *p* given the letter *a* in the word *apple*
- Similar to the first point, *p* should be likely in the context of *ap*
- The *letter l* should also be likely given the context of *app*
- The *letter e* should be likely given the context of *appl*

We start to encode each character in the word *apple* into a vector using 1-of-k encoding. 1-of-k encoding represents each character in the word as all zeros, except for the single 1 at the index of the character in the vocabulary. Each character thus represented with 1-of-k encoding is then fed into the RNN one at a time with the help of a step function. The RNN takes this input and generates a four-dimensional output vectors (one dimension per character, and recollect we only have four characters in our vocabulary). This output vector can be interpreted as the confidence that the RNN currently assigns to each character coming next in the sequence. The following diagram is a visualization of the RNN learning the characters:

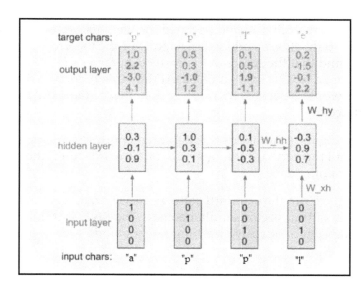

RNN learning the character language model

In the preceding diagram, we see an RNN with four-dimensional input and output layers. There is also a hidden layer with three neurons. The diagram displays the activations in the forward pass when the RNN is fed with the input of the characters *a*, *p*, *p*, and *l*. The output layer contains the confidence that the RNN assigned to each of the following characters. The expectation of the RNN is for the green numbers in the output layer to be higher than the red numbers. The high values of green numbers enable the prediction of the right characters as per the input.

We see that in the first time step, when the RNN is fed the input character *a*, it assigned a confidence of 1.0 to the next letter being *a*, 2.2 as confidence to letter *p*, -3.0 to *l*, and 4.1 to *e*. As per our training data, the sequence we considered is *apple*; therefore, the next correct character is *p* given the character *a* as input in the first time step. We would like our RNN to maximize the confidence in the first step (indicated in green) and minimize the confidence of all other letters (indicated in red). Likewise, we have a desired output character at each one of the four time steps that we would like our RNN to assign a greater confidence to.

Since the RNN consists entirely of differentiable operations, we can run the backpropagation algorithm to figure out in what direction we should adjust each one of its weights to increase the scores of the correct targets (the bold green numbers).

Based on the gradient direction, the parameters are updated and the algorithm actually alters the weight by a tiny amount in the same direction as that of the gradient. Ideally, if gradient decent has successfully run and updated the weights, we would see a slightly higher weight for the right choice and lower weights for the incorrect characters. For example, we would find that the scores of the correct character *p* in the first time step would be slightly higher, say 2.3 instead of 2.2. At the same time, the scores for the other characters *a*, *l*, and *e* would be observed as lower than that of the score that was assigned prior to gradient descent.

The process of updating the parameters through gradient descent is repeated multiple times in the RNN until the network converges, in other words, until the predictions are consistent with the training data.

Technically speaking, we run the standard softmax classifier, otherwise called the cross-entropy loss, on every output vector simultaneously. The RNN is trained with mini-batch stochastic gradient descent or adaptive learning rate methods such as RMSProp or Adam to stabilize the updates.

You may notice that the first time the character *p* is input, the output is *p*; however, the second time the same input is fed, the output is *l*. An RNN, therefore, cannot rely only on the input that is given. This is where an RNN uses its recurrent connection to keep track of the context to perform the task and make the correct predictions. Without the context, it would have been challenging for the network to predict the right output specifically, given the same input.

When we have to generate text using the trained RNN model, we provide a seed input character to the network and get the distribution over what characters are likely to come next. The distribution is then sampled and fed it right back in, to get the next letter. The process is repeated until the maximum number of characters is reached (until a specific user-defined character length), or until the model encounters an end of line character such as <EOS> or <END>.

Implementing the project

Now that we know how an RNN is able to build a character-level model, let's implement the project to generate our own words and sentences through an RNN. Generally, RNN training is computationally intensive and it is suggested that we run the code on a **graphical processing unit (GPU)**. However, due to infrastructure limitations, we are not going to use a GPU for the project code. The `mxnet` library allows a character-level language model with an RNN to be executed on the CPU itself, so let's start coding our project:

```
# including the required libraries
library("readr")
library("stringr")
library("stringi")
library("mxnet")
library("languageR")
```

To use the `languageR` library's *ALICE'S ADVENTURES IN WONDERLAND* book text and load it into memory, use the following code:

```
data(alice)
```

Next, we transform the test into feature vectors that is fed into the RNN model. The `make_data` function reads the dataset, cleans it of any non-alphanumeric characters, splits it into individual characters and groups it into sequences of length `seq.len`. In this case, `seq.len` is set to `100`:

```
make_data <- function(txt, seq.len = 32, dic=NULL) {
  text_vec <- as.character(txt)
```

```
    text_vec <- stri_enc_toascii(str = text_vec)
    text_vec <- str_replace_all(string = text_vec, pattern = "[^[:print:]]",
replacement = "")
    text_vec <- strsplit(text_vec, '') %>% unlist
    if (is.null(dic)) {
      char_keep <- sort(unique(text_vec))
    } else char_keep <- names(dic)[!dic == 0]
```

To remove those terms that are not part of dictionary, use the following code:

```
text_vec <- text_vec[text_vec %in% char_keep]
```

To build a dictionary and adjust it by −1 to have a 1-lag for labels, use the following code:

```
dic <- 1:length(char_keep)
 names(dic) <- char_keep
 # reversing the dictionary
 rev_dic <- names(dic)
 names(rev_dic) <- dic
 # Adjust by −1 to have a 1-lag for labels
 num.seq <- (length(text_vec) - 1) %/% seq.len
 features <- dic[text_vec[1:(seq.len * num.seq)]]
 labels <- dic[text_vec[1:(seq.len*num.seq) + 1]]
 features_array <- array(features, dim = c(seq.len, num.seq))
 labels_array <- array(labels, dim = c(seq.len, num.seq))
 return (list(features_array = features_array, labels_array = labels_array,
dic = dic, rev_dic
 = rev_dic))
 }
```

Set the sequence length as 100, then build the long sequence of text from individual words in alice data character vector. Then call the make_data() function on the alice_in_wonderland text file. Observe that seq.ln and an empty dictionary is passed as input. seq.ln dictates the context that is the number of characters that the RNN need to look back inorder to generate the next character. During the training seq.ln is utilized to get the right weights:

```
seq.len <- 100
 alice_in_wonderland<-paste(alice,collapse=" ")
 data_prep <- make_data(alice_in_wonderland, seq.len = seq.len, dic=NULL)
```

To view the prepared data, use the following code:

```
print(str(data_prep))
```

This will give the following output:

```
> print(str(data_prep))
List of 4
 $ features_array: int [1:100, 1:1351] 9 31 25 13 17 1 45 1 9 15 ...
 $ labels_array  : int [1:100, 1:1351] 31 25 13 17 1 45 1 9 15 51 ...
 $ dic           : Named int [1:59] 1 2 3 4 5 6 7 8 9 10 ...
  ..- attr(*, "names")= chr [1:59] " " "-" "[" "]" ...
 $ rev_dic       : Named chr [1:59] " " "-" "[" "]" ...
  ..- attr(*, "names")= chr [1:59] "1" "2" "3" "4" ...
```

To view the features array, use the following code:

```
# Viewing the feature array
View(data_prep$features_array)
```

This will give the following output:

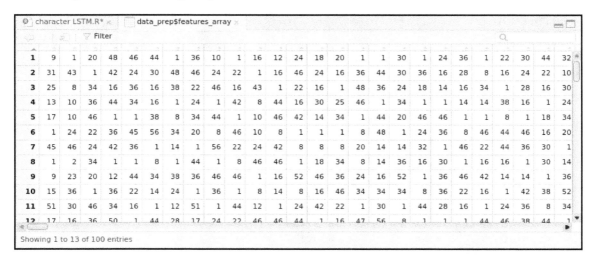

To view the labels array, use the following code:

```
# Viewing the labels array
View(data_prep$labels_array)
```

You will get the following output:

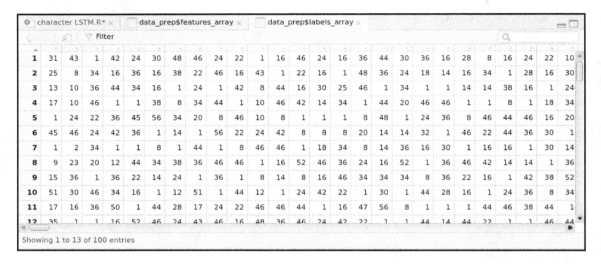

Now, let's print the dictionary, which includes the unique characters, using the following code:

```
# printing the dictionary - the unique characters
print(data_prep$dic)
```

You will get the following output:

```
> print(data_prep$dic)
    -  [  ]  *  0  3  a  A  b  B  c  C  d  D  e  E  f  F  g  G  h  H  i  I
  j  J  k  K  l  L  m  M  n  N  o  O  p
  1  2  3  4  5  6  7  8  9 10 11 12 13 14 15 16 17 18 19 20 21 22 23 24 25
 26 27 28 29 30 31 32 33 34 35 36 37 38
  P  q  Q  r  R  s  S  t  T  u  U  v  V  w  W  x  X  y  Y  z  Z
 39 40 41 42 43 44 45 46 47 48 49 50 51 52 53 54 55 56 57 58 59
```

Use the following code to print the indexes of the characters:

```
# printing the indexes of the characters
print(data_prep$rev_dic)
```

This will give the following output:

```
    1    2    3    4    5    6    7    8    9   10   11   12   13   14   15   16   17   18   19
   20   21   22   23   24   25   26   27   28
  " " "-" "[" "]" "*" "0" "3" "a" "A" "b" "B" "c" "C" "d" "D" "e" "E" "f" "F"
  "g" "G" "h" "H" "i" "I" "j" "J" "k"
   29   30   31   32   33   34   35   36   37   38   39   40   41   42   43   44   45   46   47
```

```
48   49   50   51   52   53   54   55   56
"K"  "l"  "L"  "m"  "M"  "n"  "N"  "o"  "O"  "p"  "P"  "q"  "Q"  "r"  "R"  "s"  "S"  "t"  "T"
"u"  "U"  "v"  "V"  "w"  "W"  "x"  "X"  "y"
 57   58   59
"Y"  "z"  "Z"
```

Use the following code block to fetch the features and labels to train the model, split the data into training and evaluation in a 90:10 ratio:

```
X <- data_prep$features_array
Y <- data_prep$labels_array
dic <- data_prep$dic
rev_dic <- data_prep$rev_dic
vocab <- length(dic)
samples <- tail(dim(X), 1)
train.val.fraction <- 0.9
X.train.data <- X[, 1:as.integer(samples * train.val.fraction)]
X.val.data <- X[, -(1:as.integer(samples * train.val.fraction))]
X.train.label <- Y[, 1:as.integer(samples * train.val.fraction)]
X.val.label <- Y[, -(1:as.integer(samples * train.val.fraction))]
train_buckets <- list("100" = list(data = X.train.data, label =
X.train.label))
eval_buckets <- list("100" = list(data = X.val.data, label = X.val.label))
train_buckets <- list(buckets = train_buckets, dic = dic, rev_dic =
rev_dic)
eval_buckets <- list(buckets = eval_buckets, dic = dic, rev_dic = rev_dic)
```

Use the following code to create iterators for training and evaluation datasets:

```
vocab <- length(eval_buckets$dic)
batch.size <- 32
train.data <- mx.io.bucket.iter(buckets = train_buckets$buckets, batch.size
= batch.size, data.mask.element = 0, shuffle = TRUE)
eval.data <- mx.io.bucket.iter(buckets = eval_buckets$buckets, batch.size =
batch.size,data.mask.element = 0, shuffle = FALSE)
```

Create a multi-layer RNN model to sample from character-level language models. It has a one-to-one model configuration since, for each character, we want to predict the next one. For a sequence of length `100`, there are also `100` labels, corresponding to the same sequence of characters but offset by a position of +1. The parameter's `output_last_state` is set to `TRUE`, this is to access the state of the RNN cells when performing inference and we can see `lstm` cells are used.

```
rnn_graph_one_one <- rnn.graph(num_rnn_layer = 3,
                               num_hidden = 96,
                               input_size = vocab,
                               num_embed = 64,
```

```
                            num_decode = vocab,
                            dropout = 0.2,
                            ignore_label = 0,
                            cell_type = "lstm",
                            masking = F,
                            output_last_state = T,
                            loss_output = "softmax",
                            config = "one-to-one")
```

Use the following code to visualize the RNN model:

```
graph.viz(rnn_graph_one_one, type = "graph",
          graph.height.px = 650, shape=c(500, 500))
```

The following diagram shows the resultant output:

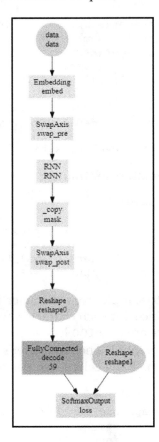

Now, use the following line of code to set the CPU as the device to execute the code:

```
devices <- mx.cpu()
```

Then, initializing the weights of the network through the Xavier initializer:

```
initializer <- mx.init.Xavier(rnd_type = "gaussian", factor_type = "avg",
magnitude = 3)
```

Use the `adadelta` optimizer to update the weights in the network through the learning process:

```
optimizer <- mx.opt.create("adadelta", rho = 0.9, eps = 1e-5, wd = 1e-8,
                           clip_gradient = 5, rescale.grad = 1/batch.size)
```

Use the following lines of code to set up logging of metrics and define a custom measurement function:

```
logger <- mx.metric.logger()
epoch.end.callback <- mx.callback.log.train.metric(period = 1, logger =
logger)
batch.end.callback <- mx.callback.log.train.metric(period = 50)
mx.metric.custom_nd <- function(name, feval) {
  init <- function() {
    c(0, 0)
  }
  update <- function(label, pred, state) {
    m <- feval(label, pred)
    state <- c(state[[1]] + 1, state[[2]] + m)
    return(state)
  }
  get <- function(state) {
    list(name=name, value = (state[[2]] / state[[1]]))
  }
  ret <- (list(init = init, update = update, get = get))
  class(ret) <- "mx.metric"
  return(ret)
}
```

Perplexity is a measure of how variable a prediction model is. If perplexity is a measure of prediction error, define a function to compute the error, using the following lines of code:

```
mx.metric.Perplexity <- mx.metric.custom_nd("Perplexity", function(label,
pred) {
  label <- mx.nd.reshape(label, shape = -1)
  label_probs <- as.array(mx.nd.choose.element.0index(pred, label))
  batch <- length(label_probs)
  NLL <- -sum(log(pmax(1e-15, as.array(label_probs)))) / batch
```

```
    Perplexity <- exp(NLL)
    return(Perplexity)
}
```

Use the following code to execute the model creation and you will see that in this project we are running it for 20 iterations:

```
model <- mx.model.buckets(symbol = rnn_graph_one_one,
                          train.data = train.data, eval.data = eval.data,
                          num.round = 20, ctx = devices, verbose = TRUE,
                          metric = mx.metric.Perplexity,
                          initializer = initializer,
         optimizer = optimizer,
                          batch.end.callback = NULL,
                          epoch.end.callback = epoch.end.callback)
```

This will give the following output:

```
Start training with 1 devices
[1] Train-Perplexity=23.490355102639
[1] Validation-Perplexity=17.6250266989171
[2] Train-Perplexity=14.4508382001841
[2] Validation-Perplexity=12.8179427398927
[3] Train-Perplexity=10.8156810097278
[3] Validation-Perplexity=9.95208184606089
[4] Train-Perplexity=8.6432934902383
[4] Validation-Perplexity=8.21806492033906
[5] Train-Perplexity=7.33073759154393
[5] Validation-Perplexity=7.03574648385079
[6] Train-Perplexity=6.32024660528852
[6] Validation-Perplexity=6.1394327776089
[7] Train-Perplexity=5.61888374338248
[7] Validation-Perplexity=5.59925324885983
[8] Train-Perplexity=5.14009899947491]
[8] Validation-Perplexity=5.29671693342219
[9] Train-Perplexity=4.77963053659987
[9] Validation-Perplexity=4.98471501141549
[10] Train-Perplexity=4.5523402301526
[10] Validation-Perplexity=4.84636357676712
[11] Train-Perplexity=4.36693337145912
[11] Validation-Perplexity=4.68806078057635
[12] Train-Perplexity=4.21294955131918
[12] Validation-Perplexity=4.53026345109037
[13] Train-Perplexity=4.08935886339982
[13] Validation-Perplexity=4.50495393289961
[14] Train-Perplexity=3.99260373800419
[14] Validation-Perplexity=4.42576079641165
[15] Train-Perplexity=3.91330125104996
```

```
[15] Validation-Perplexity=4.3941619024578
[16] Train-Perplexity=3.84730588206837
[16] Validation-Perplexity=4.33288830915229
[17] Train-Perplexity=3.78711049085869
[17] Validation-Perplexity=4.28723362252784
[18] Train-Perplexity=3.73198720637659
[18] Validation-Perplexity=4.22839393379393
[19] Train-Perplexity=3.68292148768833
[19] Validation-Perplexity=4.22187018296206
[20] Train-Perplexity=3.63728269095417
[20] Validation-Perplexity=4.17983276293299
```

Next, save the model for later use, then load the model from the disk to infer and sample the text character by character, and finally merge the predicted characters into a sentence using the following code:

```
mx.model.save(model, prefix = "one_to_one_seq_model", iteration = 20)
# the generated text is expected to be similar to the training data
set.seed(0)
model <- mx.model.load(prefix = "one_to_one_seq_model", iteration = 20)
internals <- model$symbol$get.internals()
sym_state <- internals$get.output(which(internals$outputs %in%
"RNN_state"))
sym_state_cell <- internals$get.output(which(internals$outputs %in%
"RNN_state_cell"))
sym_output <- internals$get.output(which(internals$outputs %in%
"loss_output"))
symbol <- mx.symbol.Group(sym_output, sym_state, sym_state_cell)
```

Use the following code to provide the seed character to start the text with:

```
infer_raw <- c("e")
infer_split <- dic[strsplit(infer_raw, '') %>% unlist]
infer_length <- length(infer_split)
infer.data <- mx.io.arrayiter(data = matrix(infer_split), label =
matrix(infer_split), batch.size = 1, shuffle = FALSE)
infer <- mx.infer.rnn.one(infer.data = infer.data,
                          symbol = symbol,
                          arg.params = model$arg.params,
                          aux.params = model$aux.params,
                          input.params = NULL,
                          ctx = devices)
pred_prob <- as.numeric(as.array(mx.nd.slice.axis(infer$loss_output, axis =
0, begin = infer_length-1, end = infer_length)))
pred <- sample(length(pred_prob), prob = pred_prob, size = 1) - 1
predict <- c(predict, pred)
for (i in 1:200) {
  infer.data <- mx.io.arrayiter(data = as.matrix(pred), label =
```

```
    as.matrix(pred), batch.size = 1,
  shuffle = FALSE)
    infer <- mx.infer.rnn.one(infer.data = infer.data,
                                symbol = symbol,
                                arg.params = model$arg.params,
                                aux.params = model$aux.params,
                                input.params = list(rnn.state = infer[[2]],
                                rnn.state.cell = infer[[3]]),
                                ctx = devices)
    pred_prob <- as.numeric(as.array(infer$loss_output))
    pred <- sample(length(pred_prob), prob = pred_prob, size = 1, replace =
  T) - 1
    predict <- c(predict, pred)
  }
```

Use the following lines of code to print the predicted text, after processing the predicted characters and merging them together into one sentence:

```
predict_txt <- paste0(rev_dic[as.character(predict)], collapse = "")
predict_txt_tot <- paste0(infer_raw, predict_txt, collapse = "")
# printing the predicted text
print(predict_txt_tot)
```

This will give the following output:

```
[1] "eNAHare I eat and in Heather where and fingo I ve next feeling or
fancy to livery dust a large pived as a pockethion What isual child for of
cigstening to get in a strutching voice into saying she got reaAlice glared
in a Grottle got to sea-paticular and when she heard it would heard of
having they began whrink bark of Hearnd again said feeting and there was
going to herself up it Then does so small be THESE said Alice going my dear
her before she walked at all can t make with the players and said the
Dormouse sir your mak if she said to guesss I hadn t some of the crowd and
one arches how come one mer really of a gomoice and the loots at encand
something of one eyes purried asked to leave at she had Turtle might I d
interesting tone hurry of the game the Mouse of puppled it They much put
eagerly"
```

We see from the output that our RNN is able to autogenerate text. Of course, the generated text is not very cohesive and it needs some improvement. There are several techniques we could rely upon to improve the cohesion and generate more meaningful text from an RNN. The following are some of these techniques:

- Implement a word-level language model instead of a character-level language model.
- Use a larger RNN network.

- In our project, we used LTSM cells to build our RNN. Instead of LSTM cells, we could use GRU cells, which are more advanced.

- We ran our RNN training for 20 iterations; this may be too little to get the right weights in place. We could try increasing the number of iterations and verifying the RNN yields better predictions.

- The current model used a dropout of 20%. This can be altered to check the effect on the overall predictions.

- Our corpus retained very little punctuation; therefore, our model did not have the ability to predict punctuation as characters while generating text. Including punctuation in the corpus on which an RNN gets trained may yield better sentences and word endings.

- The `seq.ln` parameter decides the number of characters that need to be looked up in the history, prior to predicting the next character. In our model, we have set this as 100. This may be altered to check whether the model produces better words and sentences.

Due to space and time constraints, we are not going to be trying these options in this chapter. One or more of these options may be experimented with by interested readers to produce better words and sentences using a character RNN.

Summary

The major theme of this chapter was generating text automatically using RNNs. We started the chapter with a discussion about language models and their applications in the real world. We then carried out an in-depth overview of recurrent neural networks and their suitability for language model tasks. The differences between traditional feedforward networks and RNNs were discussed to get a clearer understanding of RNNs. We then went on to discuss problems and solutions related to the exploding gradients and vanishing gradients experienced by RNNs. After acquiring a detailed theoretical foundation of RNNs, we went ahead with implementing a character-level language model with an RNN. We used *Alice's Adventures in Wonderland* as a text corpus input to train the RNN model and then generated a string as output. Finally, we discussed some ideas for improving our character RNN model.

How about implementing a project to win more often when playing casino slot machines? This is something we will explore in the last but one chapter of this book. Chapter 9 is titled *Winning the Casino Slot Machine with Reinforcement Learning*. Come on, let's learn to earn free money.

9
Winning the Casino Slot Machines with Reinforcement Learning

If you have been following **machine learning** (**ML**) news, I am sure you will have encountered this kind of headline: *computers performing better than world champions in various games.* If you haven't, the following are sample news snippets from my quick Google search that are worth spending time reading to understand the situation:

- Check this out: `https://www.theverge.com/2017/10/18/16495548/deepmind-ai-go-alphago-zero-self-taught/`:

> GOOGLE \ SCIENCE \ TECH \
>
> ## DeepMind's Go-playing AI doesn't need human help to beat us anymore
>
> *The company's latest AlphaGo AI learned superhuman skills by playing itself over and over*
>
> By James Vincent | @jjvincent | Oct 18, 2017, 1:00pm EDT

- See this: `https://www.makeuseof.com/tag/ais-winning-5-times-computers-beat-humans/`:

> ### 4. Deepmind, the Self-Taught
>
> Google's Deepmind may finally give nerds something to worry about because it's beating humans at **classic Atari games** — well, certain games at least. Humanity still keeps it's edge in games like Asteroid and Gravitar.

Reinforcement learning (**RL**) is a subarea of **artificial intelligence** (**AI**) that powers computer systems who are able to demonstrate better performance in games such as Atari Breakout and Go than human players.

In this chapter, we will look at the following topics:

- The concept of RL
- The multi-arm bandit problem
- Methods for solving the multi-arm bandit problem
- Real-world applications of RL
- Implementing a project using RL techniques to maximize our chances of winning at a multi-arm bandit machine

Understanding RL

RL is a very important area but is sometimes overlooked by practitioners for solving complex, real-world problems. It is unfortunate that even most ML textbooks focus only on supervised and unsupervised learning while totally ignorning RL.

RL as an area has picked up momentum in recent years; however, its origins date back to 1980. It was invented by Rich Sutton and Andrew Barto, Rich's PhD thesis advisor. It was thought of as archaic, even back in the 1980s. Rich, however, believed in RL and its promise, maintaining that it would eventually be recognized.

A quick Google search with the term RL shows that RL methods are often used in games, such as checkers and chess. Gaming problems are problems that require taking actions over time to find a long-term optimal solution to a dynamic problem. They are dynamic in the sense that the conditions are constantly changing, sometimes in response to other agents, which can be adversarial.

Although the success of RL is proven in the area of games, it is also an emerging area that is increasingly applied in other fields, such as finance, economics, and other inter-disciplinary areas. There are a number of methods in the RL area that have grown independently within the AI and operations research communities. Therefore, it is key area for a ML practitioners to learn about.

In simple terms, RL is an area that mainly focuses on creating models that learn from mistakes. Imagine that a person is put in a new environment. At first, they will make mistakes, but they will learn from them, so that when the same situation should arise in future, they will not make the same mistake again. RL uses the same technique to train the model as follows:

Environment ----------> Try and fail -----------> Learn from failures ----------> Reach goal

Historically, you couldn't use ML to get an algorithm learn how to become better than a human at performing a certain task. All that could be done was model the machine's behavior after a human's actions and, maybe, the computer would run through them faster. RL, however, makes it possible to create models that become better at performing certain tasks than humans.

Isaac Abhadu, CEO and co-founder at SYBBIO, had this wonderful explanation on Quora detailing the working of RL compared to supervised learning. He stated that an RL framework, in a nutshell, is very similar to that of supervised learning.

Suppose we're trying to get an algorithm to excel at the game of Pong. We have input frames that we will run through a model to get it to produce some random output actions, just as we would in a supervised learning setting. The difference, however, is that in the case of RL, we ourselves do not know what the target labels are, and so we don't tell the machine what's better to do in every specific situation. Instead, we apply something called a **policy gradients** method.

So, we start with a random network and feed to it an input frame so it produces a random output action to react to that frame. This action is then sent back to the game engine, which makes it produce another frame. This loop continues over and over. The only feedback it will give is the game's scoreboard. Whenever our agent does something right – that is, it produces some successful sequence – it will get a point, generally termed as a **reward**. Whenever it produces a failing sequence, it will get a point removed—this is a **penalty**.

The ultimate goal the agent is pursuing is to keep updating its policy to get as much rewards as possible. So, over time, it will figure out how to beat a human at the game.

RL is not quick. The agent is going to lose a lot at first. But we will keep feeding it frames so it keeps producing random output actions, and it will stumble upon actions that are successful. It will keep accumulating knowledge about what moves are successful and, after a while, will become invincible.

Comparison of RL with other ML algorithms

RL involves an **environment**, which is the problem set to be solved, and an **agent**, which is simply the AI algorithm. The agent will perform a certain action and the result of the action will be a change in the **state** of the agent. The change leads to the agent getting a reward, which is a positive reward, or a penalty, which is a negative reward for having performed an incorrect action. By repeating the action and reward process, the agent learns the environment. It understands the various states and the various actions that are desirable and undesirable. This process of performing actions and learning from the rewards is RL. The following diagram is an illustration showing the relationship between the agent and the environment in RL:

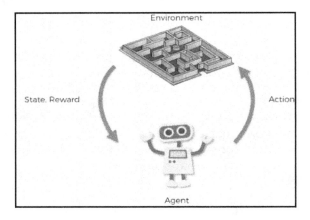

Relationship between the agent and environment in RL

RL, **deep learning** (**DL**), and ML all support automation in one way or another. All of them involve some kind of learning from the given data. However, what separates RL from the others is that RL learns the right actions by trail and error, whereas the others are focused on learning by finding patterns in the existing data. Another key difference is that for DL and ML algorithms to learn better, we will need to give them large labeled datasets, whereas this is not the case with RL.

Let's understand RL better by taking the analogy of training pets at home. Imagine we are teaching our pet dog, Santy, some new tricks. Santy, unfortunately, does not understand English; therefore, we need to find an alternative way to train him. We emulate a situation, and Santy tries to respond in many different ways. We reward Santy with a bone treat for any desirable responses. What this inculcates in the pet dog is that the next time he encounters a similar situation, he will perform the desired behavior as he knows that there is a reward. So, this is learning from positive responses; if he is treated with negative responses, such as frowning, he will be discouraged from undesirable behavior.

Terminology of RL

Let's understand the RL key terms—agent, environment, state, policy, reward, and penalty—with our pet dog training analogy:

- Our pet dog, Santy, is the agent that is exposed to the environment.
- The environment is a house or play area, depending on what we want to teach to Santy.
- Each situation encountered is called the state. For example, Santy crawling under the bed or running can be interpreted as states.
- Santy, the agent, reacts by performing actions to change from one state to another.
- After changes in states, we give the agent either a reward or a penalty, depending on the action that is performed.
- The policy refers to the strategy of choosing an action for finding better outcomes.

Now that we understand each of the RL terms, let's define the terms more formally and visualize the agent's behavior in the diagram that follows:

- **States**: The complete description of the world is known as the states. We do not abstract any information that is present in the world. States can be a position, a constant, or a dynamic. States are generally recorded in arrays, matrices, or higher order tensors.
- **Actions**: The environment generally defines the possible actions; that is, different environments lead to different actions, based on the agent. The valid actions for an agent are recorded in a space called an action space. The possible valid actions in an environment are finite in number.
- **Environment**: This is the space where the agent lives and with which the agent interacts. For different types of environments, we use different rewards and policies.
- **Reward and return**: The reward function is the one that must be kept track of at all times in RL. It plays a vital role in tuning, optimizing the algorithm, and stopping the training of the algorithm. The reward is computed based on the current state of the world, the action just taken, and the next state of the world.
- **Policies**: A policy in RL is a rule that's used by an agent for choosing the next action; the policy is also known as the agent's brain.

Take a look at the following flowchart to understand the process better:

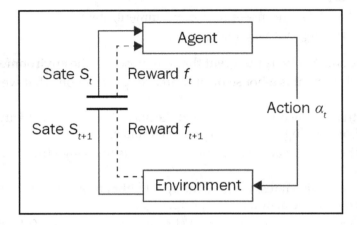

Agent behavior in RL

At each step, *t*, the agent performs the following tasks:

1. Executes action a_t
2. Receives observation s_t
3. Receives scalar reward r_t

The environment implements the following tasks:

1. Changes upon action a_t
2. Emits observation s_{t+1}
3. Emits scalar reward r_{t+1}

Time step *t* is incremented after each iteration.

The multi-arm bandit problem

Let me start with an analogy to understand this topic better. Do you like pizza? I like it a lot! My favorite restaurant in Bangalore serves delicious pizzas. I go to this place almost every time I feel like eating a pizza, and I am almost sure that I will get the best pizza. However, going to the same restaurant every time worries me that I am missing out on pizzas that are even better and served elsewhere in the town!

One alternative available is to try out restaurants one by one and sample the pizzas there, but this means that there is a very high probability that I will end up eating pizzas that aren't very nice. However, this is the one way for me to find a restaurant that serves better pizzas than the one I am currently aware of. I am aware you must be wondering why am I talking about pizzas when I am supposed to be talking about RL. Let me get to the point.

The dilemma with this task arises from incomplete information. In other words, to solve this task, it is essential to gather enough information to formulate the best overall strategy and then explore new actions. This will eventually lead to a minimization of overall bad experiences. This situation can otherwise be termed as the **exploration** versus **exploitation** dilemma:

Exploration versus exploitation dilemma

The preceding diagram aptly summarizes my best-pizza problem.

The **multi-arm bandit problem** (**MABP**) is a simplified form of the pizza analogy. It is used to represent similar kinds of problems, and finding a good strategy to solve them is already helping a lot of industries.

A **bandit** is defined as someone who steals your money! A one-armed bandit is a simple slot machine. We find this sort of machine in a casino: you insert a coin into the slot machine, pull a lever, and pray to the luck god to get an immediate reward. But the million-dollar question is why is a slot machine called a bandit? It turns out that all casinos configure the slot machines in such a way that all gamblers end up losing money!

A multi-arm bandit is a hypothetical but complicated slot machine where we have more than one slot machine lined up in a row. A gambler can pull several levers, with each lever giving a different return. The following diagram depicts the probability distribution for the corresponding reward that is different to each layer and unknown to the gambler:

Multi-arm bandit

Given these slot machines and after a set of initial trials, the task is to identify what lever to pull to get the maximum reward. In other words, pulling any one of the arms gives us a stochastic reward of either R=+1 for success, or R=0 for failure; this is called an **immediate reward**. A multi-arm bandit that issues a reward of 1 or 0 is called a **Bernoulli**. The objective is to pull the arms one-by-one in a sequence while gathering information to maximize the total payout over the long run. Formally, a Bernoulli MABP can be described as a tuple of (A,R), where the following applies:

- We have KK machines with reward probabilities, $\{\theta1,...,\theta K\}$.
- At each time step, t, we take an action, a, on one slot machine and receive a reward, r.
- A is a set of actions, each referring to the interaction with one slot machine. The value of action a is the expected reward, $Q(a) = E[r|a] = \theta$. If action a at time step t is on the i-th machine, then $Q(a_t) = \theta_i$. Q(a) is generally referred to as the action-value function.
- R is a reward function. In the case of the Bernoulli bandit, we observe a reward, r, in a stochastic fashion. At time step t, $r_t = R(a_t)$ may return reward 1 with a probability of $Q(a_t)$, or 0 otherwise.

We can solve the MABP with multiple strategies. We will review some of the strategies shortly in this section. To decide on the best strategy and to compare the different strategies, we need a quantitative method. One method is to directly compute the cumulative rewards after a certain predefined number of trials. Comparing the cumulative rewards from each of the strategies gives us an opportunity to identify the best strategies for the problem.

At times, we may already know the best action for the given bandit problem. In those cases, it may be interesting to look at the concept of regret.

Let's imagine that we know of the details of the best arm to pull for the given bandit problem. Assume that by repeatedly pulling this best arm, we get a maximum expected reward, which is shown as a horizontal line in the following diagram:

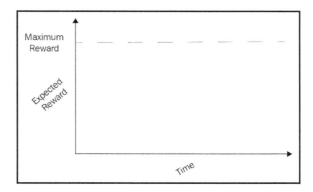

Maximum reward obtained by pulling the best arm for a MABP

As per the problem statement, we need to make repeated trials by pulling different arms of the multi-arm bandit until we are approximately sure of the arm to pull for the maximum average return at time *t*. There are a number of rounds involved while we explore and decide upon the best arm. The number of rounds, otherwise called **trials**, also incurs some loss, and this is called **regret**. In other words, we want to maximize the reward even during the learning phase. Regret can be summarized as a quantification of exactly how much we regret not picking the optimal arm.

The following diagram is an illustration showing the regret due to trials done to find the best arm:

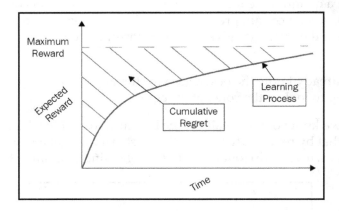

Concept of regret in MAB

Strategies for solving MABP

Based on how the exploration is done, the strategies to solve the MABP can be classified into the following types:

- No exploration
- Exploration at random
- Exploration smartly with preference to uncertainty

Let's delve into the details of some of the algorithms that fall under each of the strategy types.

Let's consider one very naive approach that involves playing just one slot machine for a long time. Here, we do no exploration at all and just randomly pick one arm to repeatedly pull to maximize the long-term rewards. You must be wondering how this works! Let's explore.

In probability theory, the law of large numbers is a theorem that describes the result of performing the same experiment a large number of times. According to this law, the average of the results obtained from a large number of trials should be close to the expected value, and will tend to become closer as more trials are performed.

We can just play with one machine for a large number of rounds so as to eventually estimate the true reward probability according to the law of large numbers.

However, there are some problems with this strategy. First and foremost, we do not know the value of a large number of rounds. Second, it is super resource intensive to play the same slot repeatedly for large number of times. And, most importantly, there is no guarantee that we will obtain the best long-term reward with this strategy.

The epsilon-greedy algorithm

The greedy algorithm in RL is a complete exploitation algorithm, which does not care for exploration. Greedy algorithms always select the action with the highest estimated action value. The action value is estimated according to past experience by averaging the rewards associated with the target action that have been observed so far.

However, use of a greedy algorithm can be a smart approach if we are able to successfully estimate the action value to the expected action value; if we know the true distribution, we can just select the best actions. An epsilon-greedy algorithm is a simple combination of the greedy and random approaches.

Epsilon helps to do this estimate. It adds exploration as part of the greedy algorithm. In order to counter the logic of always selecting the best action, as per the estimated action value, occasionally, the epsilon probability selects a random action for the sake of exploration; the rest of the time, it behaves as the original greedy algorithm and select the best known action.

The epsilon in this algorithm is an adjustable parameter that determines the probability of taking a random, rather than principled, action. It is also possible to adjust the epsilon value during training. Generally, at the start of the training process, the epsilon value is often initialized to a large probability. As the environment is unknown, the large epsilon value encourages exploration. The value is then gradually reduced to a small constant (often set to 0.1). This will increase the rate of exploitation selection.

Due to the simplicity of the algorithm, the approach has become the de facto technique for most recent RL algorithms.
Despite the common usage that the algorithm enjoys, this method is far from optimal, since it takes into account only whether actions are most rewarding or not.

Boltzmann or softmax exploration

Boltzmann exploration is also called **softmax exploration**. As opposed to either taking the optimal action all the time or taking a random action all the time, this exploration favors both through weighted probabilities. This is done through a softmax over the network's estimates of values for each action. In this case, although not guaranteed, the action that the agent estimates to be optimal is most likely to be chosen.

Boltzmann exploration has the biggest advantage over epsilon greedy. This method has information about the likely values of the other actions. In other words, let's imagine that there are five actions available to an agent. Generally, in the epsilon-greedy method, four actions are estimated as non-optimal and they are all considered equally. However, in Boltzmann exploration, the four sub-optimal choices are weighed by their relative value. This enables the agent to ignore actions that are estimated to be largely sub-optimal and give more attention to potentially promising, but not necessarily ideal, actions.

The temperature parameter (τ) controls the spread of the softmax distribution, so that all actions are considered equally at the start of training, and actions are sparsely distributed by the end of training. The parameter is annealed over time.

Decayed epsilon greedy

The value of epsilon is key in determining how well the epsilon-greedy algorithm works for a given problem. Instead of setting this value at the start and then decreasing it, we can make epsilon dependent on time. For example, epsilon can be kept equal to $1 / \log(t + 0.00001)$. As time passes, the epsilon value will keep reducing. This method works as over the time that epsilon is reduced, we become more confident of the optimal action and less exploring is required.

The problem with the random selection of actions is that after sufficient time steps, even if we know that some arm is bad, this algorithm will keep choosing that with probability *epsilon/n*. Essentially, we are exploring a bad action, which does not sound very efficient. The approach to get around this could be to favor exploration of arms with strong potential in order to get an optimal value.

The upper confidence bound algorithm

The **upper confidence bound** (**UCB**) algorithm is the most popular and widely used solution for MABPs. This algorithm is based on the principle of optimism in the face of uncertainty. This essentially means, the less uncertain we are about an arm, the more important it becomes to explore that arm.

Assume that we have two arms that can be tried out. If we have tried out the first arm 100 times but the second arm only once, then we are probably reasonably confident about the payoff of the first arm. However, we are very uncertain about the payoff of the second arm. This gives rise to the family of UCB algorithms. This can be further explained through the following diagram:

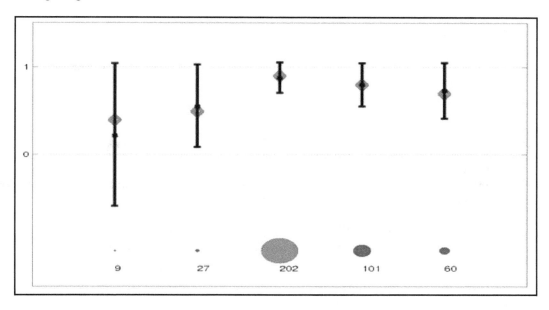

Illustration to explain upper confident bound algorithm

In the preceding diagram, each bar represents a different arm or an action. The red dot is the true expected reward and the center of the bar represents the observed average reward. The width of the bar represents the confidence interval. We are already aware that, by the law of large numbers, the more samples we have, the closer the observed average gets to the true average, and the more the bar shrinks.

The idea behind UCB algorithms is to always pick the arm or action with the highest upper bound, which is the sum of the observed average and the one-sided width of the confidence interval. This balances the exploration of arms that have not been tried many times with the exploitation of arms that have.

Thompson sampling

Thompson sampling is one of the oldest heuristics for MABPs. It is a randomized algorithm based on Bayesian ideas, and has recently generated significant interest after several studies demonstrated it to have better empirical performance compared to other methods.

There is a beautiful explanation I found on `https://stats.`
`stackexchange.com/questions/187059/could-anyone-explain-`
`thompson-sampling-in-simplest-terms`. I do not think I can do better job at explaining Thompson sampling than this. You can refer to this for further reference.

Multi-arm bandit – real-world use cases

We encounter so many situations in the real world that are similar to that of the MABP we reviewed in this chapter. We could apply RL strategies to all these situations. The following are some of the real-world use cases similar to that of the MABP:

- Finding the best medicine/s among many alternatives
- Identifying the best product to launch among possible products
- Deciding the amount of traffic (users) that we need to allocate for each website
- Identifying the best marketing strategy for launching a product
- Identifying the best stocks portfolio to maximize profit
- Finding out the best stock to invest in
- Figuring out the shortest path in a given map
- Click-through rate prediction for ads and articles
- Predicting the most beneficial content to be cached at a router based upon the content of articles
- Allocation of funding for different departments of an organization
- Picking best-performing athletes out of a group of students given limited time and an arbitrary selection threshold

So far, we have covered almost all of the basic details that we need to know to progress to the practical implementation of RL to the MABP. Let's kick-start coding solutions to the MABP in our next section.

Solving the MABP with UCB and Thompson sampling algorithms

In this project, we will use upper confidence limits and Thompson sampling algorithms to solve the MABP. We will compare their performance and strategy in three different situations—standard rewards, standard but more volatile rewards, and somewhat chaotic rewards. Let's prepare the simulation data, and once the data is prepared, we will view the simulated data using the following code:

```
# loading the required packages
library(ggplot2)
library(reshape2)
# distribution of arms or actions having normally distributed
# rewards with small variance
# The data represents a standard, ideal situation i.e.
# normally distributed rewards, well seperated from each other.
mean_reward = c(5, 7.5, 10, 12.5, 15, 17.5, 20, 22.5, 25, 26)
reward_dist = c(function(n) rnorm(n = n, mean = mean_reward[1], sd = 2.5),
                function(n) rnorm(n = n, mean = mean_reward[2], sd = 2.5),
                function(n) rnorm(n = n, mean = mean_reward[3], sd = 2.5),
                function(n) rnorm(n = n, mean = mean_reward[4], sd = 2.5),
                function(n) rnorm(n = n, mean = mean_reward[5], sd = 2.5),
                function(n) rnorm(n = n, mean = mean_reward[6], sd = 2.5),
                function(n) rnorm(n = n, mean = mean_reward[7], sd = 2.5),
                function(n) rnorm(n = n, mean = mean_reward[8], sd = 2.5),
                function(n) rnorm(n = n, mean = mean_reward[9], sd = 2.5),
                function(n) rnorm(n= n, mean = mean_reward[10], sd = 2.5))
#preparing simulation data
dataset = matrix(nrow = 10000, ncol = 10)
for(i in 1:10){
  dataset[, i] = reward_dist[[i]](n = 10000)
}
# assigning column names
colnames(dataset) <- 1:10
# viewing the dataset that is just created with simulated data
View(dataset)
```

This will give the following output:

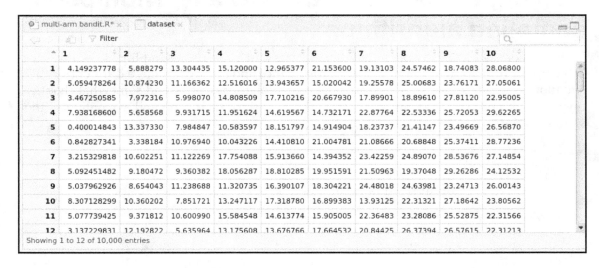

Now, create a melted dataset with an arm and reward combination, and then convert the arm column to the nominal type using the following code:

```
# creating a melted dataset with arm and reward combination
dataset_p = melt(dataset)[, 2:3]
colnames(dataset_p) <- c("Bandit", "Reward")
# converting the arms column in the dataset to nominal type
dataset_p$Bandit = as.factor(dataset_p$Bandit)
# viewing the dataset that is just melted
View(dataset_p)
```

This will give us the following output:

Now, plot the distributions of rewards from bandits using the following code:

```
#ploting the distributions of rewards from bandits
ggplot(dataset_p, aes(x = Reward, col = Bandit, fill = Bandit)) +
  geom_density(alpha = 0.3) +
  labs(title = "Reward from different bandits")
```

This will give us the following output:

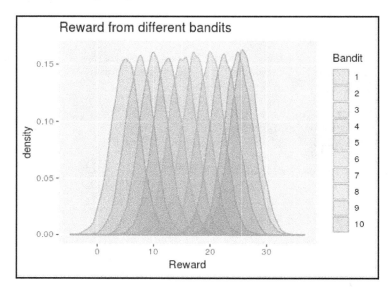

Now let's implement the UCB algorithm on the hypothesized arm with a normal distribution using the following code:

```
# implementing upper confidence bound algorithm
UCB <- function(N = 1000, reward_data){
  d = ncol(reward_data)
  bandit_selected = integer(0)
  numbers_of_selections = integer(d)
  sums_of_rewards = integer(d)
  total_reward = 0
  for (n in 1:N) {
    max_upper_bound = 0
    for (i in 1:d) {
      if (numbers_of_selections[i] > 0){
        average_reward = sums_of_rewards[i] / numbers_of_selections[i]
        delta_i = sqrt(2 * log(1 + n * log(n)^2) /
numbers_of_selections[i])
        upper_bound = average_reward + delta_i
      } else {
        upper_bound = 1e400
      }
      if (upper_bound > max_upper_bound){
        max_upper_bound = upper_bound
        bandit = i
      }
    }
    bandit_selected = append(bandit_selected, bandit)
    numbers_of_selections[bandit] = numbers_of_selections[bandit] + 1
    reward = reward_data[n, bandit]
    sums_of_rewards[bandit] = sums_of_rewards[bandit] + reward
    total_reward = total_reward + reward
  }
  return(list(total_reward = total_reward, bandit_selected bandit_selected,
numbers_of_selections = numbers_of_selections, sums_of_rewards =
sums_of_rewards))
}
# running the UCB algorithm on our
# hypothesized arms with normal distributions
UCB(N = 1000, reward_data = dataset)
```

You will get the following as the resultant output:

```
$total_reward
      1
25836.91
$numbers_of_selections
 [1]   1   1   1   1   1   1   2   1  23 968
$sums_of_rewards
```

```
[1]      4.149238     10.874230     5.998070     11.951624     18.151797
21.004781     44.266832     19.370479     563.001692
[10] 25138.139942
```

Next, we will implement the Thompson sampling algorithm using a **normal-gamma** prior and normal likelihood to estimate posterior distributions using the following code:

```
# Thompson sampling algorithm
rnormgamma <- function(n, mu, lambda, alpha, beta){
  if(length(n) > 1)
    n <- length(n)
  tau <- rgamma(n, alpha, beta)
  x <- rnorm(n, mu, 1 / (lambda * tau))
  data.frame(tau = tau, x = x)
}
T.samp <- function(N = 500, reward_data, mu0 = 0, v = 1, alpha = 2,
beta = 6){
  d = ncol(reward_data)
  bandit_selected = integer(0)
  numbers_of_selections = integer(d)
  sums_of_rewards = integer(d)
  total_reward = 0
  reward_history = vector("list", d)
  for (n in 1:N){
    max_random = -1e400
    for (i in 1:d){
      if(numbers_of_selections[i] >= 1){
        rand = rnormgamma(1,
                    (v * mu0 + numbers_of_selections[i] *
mean(reward_history[[i]])) / (v + numbers_of_selections[i]),
                    v + numbers_of_selections[i],
                    alpha + numbers_of_selections[i] / 2,
                    beta + (sum(reward_history[[i]] -
mean(reward_history[[i]])) ^ 2) / 2 + ((numbers_of_selections[i] * v) / (v
+ numbers_of_selections[i])) * (mean(reward_history[[i]]) - mu0) ^ 2 / 2)$x
      }else {
        rand = rnormgamma(1, mu0, v, alpha, beta)$x
      }
      if(rand > max_random){
        max_random = rand
        bandit = i
      }
    }
    bandit_selected = append(bandit_selected, bandit)
    numbers_of_selections[bandit] = numbers_of_selections[bandit] + 1
    reward = reward_data[n, bandit]
    sums_of_rewards[bandit] = sums_of_rewards[bandit] + reward
    total_reward = total_reward + reward
```

```
    reward_history[[bandit]] = append(reward_history[[bandit]], reward)
  }
  return(list(total_reward = total_reward, bandit_selected =
bandit_selected, numbers_of_selections = numbers_of_selections,
sums_of_rewards = sums_of_rewards))
}
# Applying Thompson sampling using normal-gamma prior and Normal likelihood
to estimate posterior distributions
T.samp(N = 1000, reward_data = dataset, mu0 = 40)
```

You will get the following as the resultant output:

```
$total_reward
      10
24434.24
$numbers_of_selections
 [1]  16  15  15  14  14  17  16  19  29 845
$sums_of_rewards
 [1]    80.22713   110.09657   141.14346   171.41301   212.86899
293.30138   311.12230   423.93256   713.54105 21976.59855
```

From the results, we can infer that the UCB algorithm quickly identified that the 10th arm yields the most reward. We also observe that Thompson sampling tried the worst arms a lot more times before finding the best one.

Now, let's simulate the data of bandits with normally distributed rewards with large variance and plot the distributions of rewards by using the following code:

```
# Distribution of bandits / actions having normally distributed rewards
with large variance
# This data represents an ideal but more unstable situation: normally
distributed rewards with much larger variance,
# thus not well separated from each other.
mean_reward = c(5, 7.5, 10, 12.5, 15, 17.5, 20, 22.5, 25, 26)
reward_dist = c(function(n) rnorm(n = n, mean = mean_reward[1], sd = 20),
                function(n) rnorm(n = n, mean = mean_reward[2], sd = 20),
                function(n) rnorm(n = n, mean = mean_reward[3], sd = 20),
                function(n) rnorm(n = n, mean = mean_reward[4], sd = 20),
                function(n) rnorm(n = n, mean = mean_reward[5], sd = 20),
                function(n) rnorm(n = n, mean = mean_reward[6], sd = 20),
                function(n) rnorm(n = n, mean = mean_reward[7], sd = 20),]
                function(n) rnorm(n = n, mean = mean_reward[8], sd = 20),
                function(n) rnorm(n = n, mean = mean_reward[9], sd = 20),
                function(n) rnorm(n = n, mean = mean_reward[10], sd = 20))
#preparing simulation data
dataset = matrix(nrow = 10000, ncol = 10)
for(i in 1:10){
  dataset[, i] = reward_dist[[i]](n = 10000)
```

```
}
colnames(dataset) <- 1:10
dataset_p = melt(dataset)[, 2:3]
colnames(dataset_p) <- c("Bandit", "Reward")
dataset_p$Bandit = as.factor(dataset_p$Bandit)
#plotting the distributions of rewards from bandits
ggplot(dataset_p, aes(x = Reward, col = Bandit, fill = Bandit)) +
  geom_density(alpha = 0.3) +
  labs(title = "Reward from different bandits")
```

You will get the following graph as the resultant output:

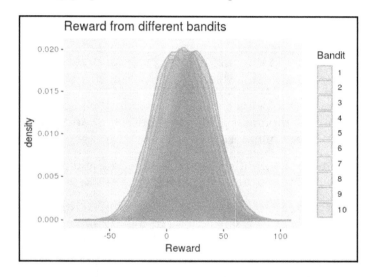

Apply UCB on rewards with higher variance using the following code:

```
# Applying UCB on rewards with higher variance
UCB(N = 1000, reward_data = dataset)
```

You will get the following output:

```
$total_reward
       1
25321.39
$numbers_of_selections
 [1]   1   1   1   3   1   1   2   6 903  81
$sums_of_rewards
 [1]     2.309649    -6.982907   -24.654597     49.186498      8.367174
-16.211632    31.243270   104.190075 23559.216706   1614.725305
```

Next, apply Thompson sampling on rewards with higher variance by using the following code:

```
# Applying Thompson sampling on rewards with higher variance
T.samp(N = 1000, reward_data = dataset, mu0 = 40)
```

You will get the following output:

```
$total_reward
      2
24120.94
$numbers_of_selections
 [1]   16   15   14   15   15   17   20   21  849   18
$sums_of_rewards
 [1]      94.27878     81.42390    212.00717    181.46489    140.43908
249.82014   368.52864   397.07629 22090.20740 305.69191
```

From the results, we can infer that when the fluctuation of rewards is greater, the UCB algorithm is more susceptible to being stuck at a suboptimal choice and never finds the optimal bandit. Thompson sampling is generally more robust and is able to find the optimal bandit in all kinds of situations.

Now let's simulate the more chaotic distribution bandit data and plot the distribution of rewards from bandits by using the following code:

```
# Distribution of bandits / actions with rewards of different distributions
# This data represents an more chaotic (possibly more realistic) situation:
# rewards with different distribution and different variance.
mean_reward = c(5, 7.5, 10, 12.5, 15, 17.5, 20, 22.5, 25, 26)
reward_dist = c(function(n) rnorm(n = n, mean = mean_reward[1], sd = 20),
                function(n) rgamma(n = n, shape = mean_reward[2] / 2, rate
                = 0.5),
                function(n) rpois(n = n, lambda = mean_reward[3]),
                function(n) runif(n = n, min = mean_reward[4] - 20, max =
mean_reward[4] + 20),
                function(n) rlnorm(n = n, meanlog = log(mean_reward[5]) -
0.25, sdlog = 0.5),
                function(n) rnorm(n = n, mean = mean_reward[6], sd = 20),
                function(n) rexp(n = n, rate = 1 / mean_reward[7]),
                function(n) rbinom(n = n, size = mean_reward[8] / 0.5, prob
= 0.5),
                function(n) rnorm(n = n, mean = mean_reward[9], sd = 20),
                function(n) rnorm(n = n, mean = mean_reward[10], sd = 20))
#preparing simulation data
dataset = matrix(nrow = 10000, ncol = 10)
for(i in 1:10){
  dataset[, i] = reward_dist[[i]](n = 10000)
}
```

```
colnames(dataset) <- 1:10
dataset_p = melt(dataset)[, 2:3]
colnames(dataset_p) <- c("Bandit", "Reward")
dataset_p$Bandit = as.factor(dataset_p$Bandit)
#plotting the distributions of rewards from bandits
ggplot(dataset_p, aes(x = Reward, col = Bandit, fill = Bandit)) +
  geom_density(alpha = 0.3) +
  labs(title = "Reward from different bandits")
```

You will get the following graph as the resultant output:

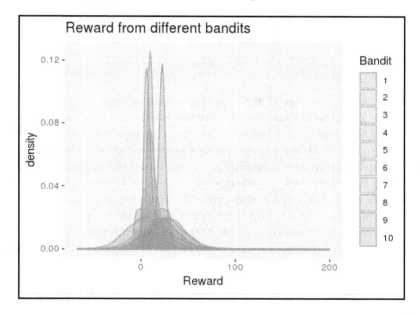

Apply UCB on rewards with different distributions by using the following code:

```
# Applying UCB on rewards with different distributions
UCB(N = 1000, reward_data = dataset)
```

You will get the following output:

```
$total_reward
        1
22254.18
$numbers_of_selections
 [1]   1   1   1   1   1   1   1 926  61   6
$sums_of_rewards
 [1]     6.810026     3.373098     8.000000    12.783859    12.858791
   11.835287     1.616978 20755.000000 1324.564987   117.335467
```

Next, apply Thompson sampling on rewards with different distributions by using the following code:

```
# Applying Thompson sampling on rewards with different distributions
T.samp(N = 1000, reward_data = dataset, mu0 = 40)
```

You will get the following as the resultant output:

```
$total_reward
      2
24014.36
$numbers_of_selections
 [1]   16   14   14   14   14   15   14   51  214  634
$sums_of_rewards
 [1]    44.37095   127.57153   128.00000   142.66207   191.44695
169.10430   150.19486   1168.00000   5201.69130  16691.32118
```

From the preceding results, we see that the performance of the two algorithms is similar. A major reason for the Thompson sampling algorithm trying all bandits several times before choosing the one it considers best is because we chose a prior distribution with a relatively high mean in this project. With the prior having a larger mean, the algorithm favors **exploration over exploitation** at the beginning. Only when the algorithm becomes very confident that it has found the best choice does it value exploitation over exploration. If we decrease the mean of the prior, exploitation would have a higher value and the algorithm would stop exploring faster. By altering the prior distribution used, you can adjust the relative importance of exploration over exploitation to suit the specific problem at hand. This is more evidence highlighting the flexibility of the Thompson sampling algorithm.

Summary

In this chapter, we learned about RL. We started the chapter by defining RL and its difference when compared with other ML techniques. We then reviewed the details of the MABP and looked at the various strategies that can be used to solve this problem. Use cases that are similar to the MABP were discussed. Finally, a project was implemented with UCB and Thompson sampling algorithms to solve the MABP using three different simulated datasets.

We have almost reached the end of this book. The appendix of this book, *The Road Ahead*, as the name reflects, is a guidance chapter suggesting details on what's next from here to become a better R data scientist. I am super excited that I am at the last leg of this R projects journey. Are you with me on this as well?

The Road Ahead

Congratulations on reaching this point in the book! It has been an exciting journey for me, writing the various chapters and implementing the projects. I learned quite a lot as a result of undertaking this journey and I hope your experience has been the same.

Finally, here we are at the concluding chapter!

The year 2018 can be considered a boom year for data science, **machine learning (ML)**, and **artificial intelligence (AI)**. Just look at the number of start-ups with the terms ML and AI in their tag lines, the focal points for acquisitions on the part of big companies, and the topics at the biggest tech conferences. It does not take a long time for us to realize that data, ML, and AI are omnipresent, and I believe that this trend is going to continue for a few years to come. More and more industries are going to utilize ML and AI in a big way. This essentially is going to create gaps in available talent in terms of implementing ML and AI in businesses. Therefore, this is the best time to invest in learning more and more in this area. AI and ML skills, combined with business skills, will be the most highly prized within the industry.

We also need to realize that the AI and ML space is evolving rapidly – new ML algorithms, new platforms, new infrastructure, and new types of data are just a few examples. To stay relevant, the only option is to keep ourselves updated with the newest trends and techniques out there. The other characteristics that are the most sought-after are unlearning the old and learning the new, and being flexible and modest enough to say, *it is okay that I don't know, but I am open to learning.*

Finally, what is the best way to learn and stay on top of things? Of course, there is no dearth of resources—**Massive Open Online Courses (MOOCs)**, books, blogs, conferences, seminars, classes, and so on. All that is needed is time and a willingness to learn. Happy learning in order to happily excel!

Other Books You May Enjoy

If you enjoyed this book, you may be interested in these other books by Packt:

Machine Learning with R Cookbook - Second Edition

AshishSingh Bhatia, Yu-Wei, Chiu (David Chiu), Recommended for You

ISBN: 9781787284395

- Create and inspect transaction datasets and perform association analysis with the Apriori algorithm
- Visualize patterns and associations using a range of graphs and find frequent item-sets using the Eclat algorithm
- Compare differences between each regression method to discover how they solve problems
- Detect and impute missing values in air quality data
- Predict possible churn users with the classification approach
- Plot the autocorrelation function with time series analysis
- Use the Cox proportional hazards model for survival analysis
- Implement the clustering method to segment customer data
- Compress images with the dimension reduction method
- Incorporate R and Hadoop to solve machine learning problems on big data

Machine Learning Algorithms - Second Edition
Giuseppe Bonaccorso

ISBN: 9781789347999

- Study feature selection and the feature engineering process
- Assess performance and error trade-offs for linear regression
- Build a data model and understand how it works by using different types of algorithm
- Learn to tune the parameters of Support Vector Machines (SVM)
- Explore the concept of natural language processing (NLP) and recommendation systems
- Create a machine learning architecture from scratch

Leave a review - let other readers know what you think

Please share your thoughts on this book with others by leaving a review on the site that you bought it from. If you purchased the book from Amazon, please leave us an honest review on this book's Amazon page. This is vital so that other potential readers can see and use your unbiased opinion to make purchasing decisions, we can understand what our customers think about our products, and our authors can see your feedback on the title that they have worked with Packt to create. It will only take a few minutes of your time, but is valuable to other potential customers, our authors, and Packt. Thank you!

Index

implementation 68, 69
natural language processing (NLP) 23, 25
network architecture 195
non-globally-optimal solution 166
normal-gamma prior 303
null function 223

O

object-oriented programming (OOP) 21
one-class classification algorithm 219
outlier detection
  drawbacks 218
over complete autoencoders 222
overfitting
  about 29, 30, 65, 202
  avoiding, with dropout 202, 204, 205, 206, 207

P

Part-of-Speech (POS) tagger 115
partial dependency plot (PDP) 34
penalty 287
performance metrics 32
personalized content recommendation
  archiving 82, 83
policy gradients method 287
predictor variables 27
pretrained models
  computer vision, implementing 212, 213, 214,
    215, 216
  Google's Word2Vec model 24
  inception-V3 model 23
  MobileNet 24
  reference 212
  Stanford's GloVe model 24
  VCG 16 24
  VCG Face 24
pretrained Word2vec word embedding
  text sentiment classifier, building with 133, 135,
    136, 137, 138
principal component analysis (PCA) 28, 164

R

R
  autoencoders, building with H2O library 227,
    228

random forests
  attrition prediction model, implementing 70, 71
  for randomization 69
real-time mode 39
recommendation engine, categories
  about 84
  collaborative filtering 84
  content-based filtering 84
  hybrid filtering 85
recommendation engine
  fundamental aspects 82, 83
recommendation system
  building, with association-rule mining 99
  building, with item-based collaborative filtering
    (ITCF) 91, 92, 94
  building, with user-based collaborative filtering
    (UBCF) 95, 96, 99
Recommender function, parameter
  data normalization 91
  distance 91
Rectified Linear Unit (ReLU) 188
recurrent neural networks (RNNs)
  about 24, 257
  automated prose generator, building 267, 268,
    270, 271, 272, 273
  exploring 257, 258, 259, 260
  gradients, problems and solutions 265
  text, generating 282
  versus feedforward neural network 261, 262,
    263
regression 16
regret 293
reinforcement learning (RL), terminology
  about 290
  action 289
  agent 289
  environment 289
  penalty 289
  policy 289
  return 289
  reward 289
  state 289
reinforcement learning (RL)
  about 19, 20, 286, 287
  comparing, with other ML algorithms 288